Br　　　　　e
Fortress Wall

Understanding Terrorist Efforts to Overcome Defensive Technologies

Brian A. Jackson · Peter Chalk · R. Kim Cragin

Bruce Newsome · John V. Parachini · William Rosenau

Erin M. Simpson · Melanie Sisson · Donald Temple

Prepared for the Department of Homeland Security

Homeland Security

A RAND INFRASTRUCTURE, SAFETY, AND ENVIRONMENT PROGRAM

This research was sponsored by the United States Department of Homeland Security and was conducted under the auspices of the Homeland Security Program within RAND Infrastructure, Safety, and Environment.

Library of Congress Cataloging-in-Publication Data

Breaching the fortress wall : understanding terrorist efforts to overcome de
    technologies / Brian A. Jackson ... [et al.].
       p. cm.
       "MG-481."
       Includes bibliographical references.
       ISBN 0-8330-3914-8 (pbk. : alk. paper)
       1. War on Terrorism, 2001-—Technology. 2. Security systems. 3. Terrorism—
Prevention. 4. Terrorism—Prevention—Case studies. 5. Terrorism—Case studies.
       6. National security. I. Jackson, Brian A. (Brian Anthony)

    HV6431.B737 2007
    363.325'72—dc22

                                                                        2006001721

The RAND Corporation is a nonprofit research organization providing objective analysis and effective solutions that address the challenges facing the public and private sectors around the world. RAND's publications do not necessarily reflect the opinions of its research clients and sponsors.

RAND® is a registered trademark.

Cover design by Eileen Delson La Russo
Photo by TSgt Cedric H. Rudisill, U.S. Air Force

Published 2007 by the RAND Corporation
1776 Main Street, P.O. Box 2138, Santa Monica, CA 90407-2138
1200 South Hayes Street, Arlington, VA 22202-5050
4570 Fifth Avenue, Suite 600, Pittsburgh, PA 15213-2665
RAND URL: http://www.rand.org/
To order RAND documents or to obtain additional information, contact
Distribution Services: Telephone: (310) 451-7002;
Fax: (310) 451-6915; Email: order@rand.org

Preface

Technical countermeasures are key components of national efforts to combat terrorist violence. Efforts to collect data about and disrupt terrorist activities through human intelligence and direct action, information gathering, and protective technologies complement technical countermeasures, helping to ensure that terrorists are identified, their ability to plan and stage attacks is limited, and, if those attacks occur, their impact is contained.

Given the potential effect of such measures on the terrorists' capabilities, it is not surprising that they act to reduce or neutralize the impact of defensive technologies on their activities. In the event that the terrorists' counterefforts are successful, the value and protection provided by defensive technologies can be substantially reduced. Through case studies of terrorist struggles in a number of nations, this document analyzes the nature and impact of such terrorist counterefforts on the value of defensive technologies deployed against them.

The information presented here should be of interest to homeland security policymakers in that it identifies potential weaknesses in defensive technology systems, thereby informing threat assessment and providing a basis for improving the design of future defensive technologies. It extends the RAND Corporation's ongoing research on terrorism and domestic security issues. Related RAND publications include the following:

- Brian A. Jackson, John C. Baker, Peter Chalk, Kim Cragin, John V. Parachini, and Horacio R. Trujillo, *Aptitude for Destruction*,

Vol. 1: *Organizational Learning in Terrorist Groups and Its Implications for Combating Terrorism*, MG-331-NIJ, 2005.
- Brian A. Jackson, John C. Baker, Peter Chalk, Kim Cragin, John V. Parachini, and Horacio R. Trujillo, *Aptitude for Destruction*, Vol. 2: *Case Studies of Organizational Learning in Five Terrorist Groups*, MG-332-NIJ, 2005.
- Kim Cragin and Sara A. Daly, *The Dynamic Terrorist Threat: An Assessment of Group Motivations and Capabilities in a Changing World*, MR-1782-AF, 2004.
- Peter Chalk and William Rosenau, *Confronting "the Enemy Within": Security Intelligence, the Police, and Counterterrorism in Four Democracies*, MG-100-RC, 2004.

This monograph is one in a series of studies examining technological issues in terrorism and efforts to combat it. This series focuses on understanding how terrorist groups make technology choices and respond to the technologies deployed against them. This research was sponsored by the U.S. Department of Homeland Security, Science and Technology Directorate, Office of Comparative Studies.

The RAND Homeland Security Program

This research was conducted under the auspices of the Homeland Security Program within RAND Infrastructure, Safety, and Environment (ISE). The mission of RAND Infrastructure, Safety, and Environment is to improve the development, operation, use, and protection of society's essential physical assets and natural resources and to enhance the related social assets of safety and security of individuals in transit and in their workplaces and communities. Homeland Security Program research supports the Department of Homeland Security and other agencies charged with preventing and mitigating the effects of terrorist activity within U.S. borders. Projects address critical infrastructure protection, emergency management, terrorism risk management, border control, first responders and preparedness, domestic threat assessments, domestic intelligence, and workforce and training.

Questions or comments about this monograph should be sent to the project leader, Brian A. Jackson (Brian_Jackson@rand.org). Information about the Homeland Security Program is available online (http://www.rand.org/ise/security/). Inquiries about homeland security research projects should be sent to the following address:

Michael Wermuth, Director
Homeland Security Program, ISE
RAND Corporation
1200 South Hayes Street
Arlington, VA 22202-5050
703-413-1100, x5414
Michael_Wermuth@rand.org

Contents

Preface iii
Figures xi
Tables xiii
Summary xv
Acknowledgments xxv
Abbreviations xxvii

CHAPTER ONE

Introduction 1
Defensive Technologies and the Effort to Combat Terrorism 3
Terrorist Efforts to Overcome Defensive Technologies 7
About the Study 8
About This Monograph 11

CHAPTER TWO

Palestinian Terrorist Groups 13
Introduction 13
Information Acquisition and Management 22
Preventive Action 26
Denial 29
Response 34
Conclusion 34

CHAPTER THREE
Jemaah Islamiyah and Affiliated Groups. 39
Introduction 39
Information Acquisition and Management 42
Preventive Action. 48
Denial 52
Investigation 54
Conclusion 55

CHAPTER FOUR
Liberation Tigers of Tamil Eelam 59
Introduction 59
Information Acquisition and Management 69
Denial 76
Conclusion 79

CHAPTER FIVE
Provisional Irish Republican Army 83
Introduction 83
Information Acquisition and Management 85
Preventive Action. 96
Denial 99
Response 103
Investigation 105
Conclusion 109

CHAPTER SIX
Conclusions:
   Understanding Terrorists' Countertechnology Efforts. 115
Terrorist Strategies for Countering Defensive Technologies 116
Transferability of Terrorist Countertechnology Strategies 121
Implications of Terrorist Countertechnology Activities for
   Homeland Security Efforts. 125
In Conclusion: The Role of Technology in Combating Terrorism 132

APPENDIX
Prominent Acts of LTTE Suicide Terrorism, 1987–2002 135

Bibliography 139

Figures

1.1. The Terrorist Activity Chain 2
1.2. Defensive Technologies Across the Terrorist Activity Chain 5
2.1. Terrorist Attacks in Israel, the West Bank, and Gaza Strip
     During the Oslo Period, 1993–1999 15
2.2. Terrorist Attacks During the al-Aqsa Intifada, 2000–2005 16
2.3. Distribution of Terrorist Attacks Since September 2000 18
2.4. Route Map of Israeli Security Barrier 31
6.1. Terrorist Countertechnology Strategies 119

Tables

2.1. Palestinian Groups' Technological Innovations: Purpose and Intended Mitigation of Government Countermeasures 35

3.1. Jemaah Islamiyah Technological Innovations: Purpose and Intended Mitigation of Government Countermeasures 55

4.1. Estimates of the Efficiency and Effectiveness Matrix of Search Modalities According to Sri Lankan Intelligence Sources 71

4.2. Liberation Tigers of Tamil Eelam Technological Innovations: Purpose and Intended Mitigation of Government Countermeasures 80

5.1. Provisional Irish Republican Army Technological Innovations: Purpose and Intended Mitigation of Government Countermeasures 109

A.1. Prominent Acts of LTTE Suicide Terrorism, 1987–2002 135

Summary

The level of threat posed by a terrorist group[1] is determined in large part by its ability to build its organizational capabilities and bring those capabilities to bear in violent action. As part of homeland security efforts, technology systems play a key role within a larger, integrated strategy to target groups' efforts to do so and protect the public from the threat of terrorist violence. Although many types of technology have roles to play in the overall effort to fight terrorism, this analysis focuses on a class of tools that we call *defensive technologies*—the systems and approaches deployed to protect an area and its citizens from terrorism by discovering and frustrating the plans of terrorists operating therein. The technologies that we have defined as defensive technologies can be organized into five primary classes based on their intended impact on the terrorist adversary:

- **Information acquisition and management.** These tools include surveillance technologies and practices that enable law enforce-

---

[1] Although some of the substate groups discussed in this book use tactics that are not purely terroristic in nature—for example, mixing traditional military operations against opposing security forces with terrorist bombings or assassinations—we use the terms *terrorism* and *terrorist violence* as generic descriptors of the violent activities of substate groups.

In this book, we adopt the convention that *terrorism* is a tactic—the systematic and premeditated use, or threatened use, of violence by nonstate groups to further political or social objectives to coerce an audience larger than those directly affected. With terrorism defined as a tactic, it follows that individual organizations are not inherently *terrorist*. We use the terms *terrorist group* and *terrorist organization* as shorthand for "group that has chosen to use terrorism."

ment and security organizations to gather information on terrorist individuals, vehicles, and behaviors; to monitor sites and areas (including border information systems aimed at excluding terrorists from the country); to detect concealed weapons and operations in progress; and to maintain the profiles, databases, and systems needed to manage and use such information once collected.

- **Preventive action.** Technologies in this category include systems to counter specific terrorist weapon systems (e.g., radio-detonator jamming, antimissile systems) and systems designed to prevent terrorist access to money, weapons, technologies, and other resources or knowledge.

- **Denial.** Such approaches include traditional hardening of potential targets (e.g., setbacks, blast walls, reinforced windows, or other structures); design changes in potential targets to make them less susceptible to attack (e.g., increasing the robustness of infrastructure systems, immune buildings); hardening of the population (e.g., psychological preparedness efforts, vaccination); and security or guard force deployment.

- **Response.** These technologies are designed to provide multiple capabilities, including defeating operations in progress (e.g., explosive ordnance disposal teams); ensuring that emergency responses are adequate to treat casualties and limit the spread of damage from attacks in progress; coordinating response operations for increased effectiveness; making antidotes or other treatment methods for specific types of terrorist attacks readily available; and providing risk communication capabilities, which can be used to shape public responses to minimize the effects of an attack.

- **Investigation.** Technologies in this domain include forensic science and other investigative and identification technologies to analyze terrorist weapons, track and apprehend suspects, support prosecution of individuals responsible for terrorist operations, or enable other sovereign action against individuals or terrorist organizations.

These do not represent the only technologies relevant to efforts to combat terrorism. A range of technologies applied in more proactive or offensive operations against terrorist groups—including military weaponry and similar technologies—are not included within the scope of this book. It should also be noted that the distinction between offensive and defensive technologies is admittedly ambiguous; the same intelligence-gathering system deployed in a defensive mode to detect terrorist operations in progress could clearly gather information supporting offensive operations against terrorist organizations.

Although the contributions that technology can make in combating terrorism can be considerable, it should be noted that technology is only one of many tools for combating terrorism. For example, virtually all sources consulted for this book emphasized the preeminence of direct human intelligence—through infiltration of terrorist organizations or the recruitment of their members as agents—as the most important element of an effort to combat terrorists' activities.[2] The emphasis of this book on technical systems should not be interpreted as contradicting this view—rather, the work described here should be seen as part of a multifaceted effort against terrorism to ensure that technology complements other efforts as effectively and efficiently as possible.

Terrorist Efforts to Overcome Defensive Measures

Although the variety of defensive technologies available enables broad-based targeting of terrorists' activities, defending a nation against terrorism is not a one-sided game. Given the potential for defensive technologies to constrain the capabilities of terrorist groups and limit their operational freedom, these organizations are acutely aware of government efforts to deploy them and actively seek ways to evade or counteract them. This measure-countermeasure, move-countermove dynamic is inherent in contests between organizations and, to the extent that

---

[2] Personal interviews with former law enforcement officials, England (May 2005) and with local officials, Indonesia, the Philippines, Singapore, and Thailand (March–April 2005).

the terrorists' efforts are successful, can significantly reduce or eliminate the value of defensive technologies.[3]

This book focuses on understanding terrorists' countertechnology efforts by drawing on relevant data from the history of a variety of terrorist conflicts and applying that information to the broader technological questions relevant to current homeland security efforts. These cases were selected for examination:

- **Palestinian terrorist groups.** In Israel, a variety of Palestinian terrorist groups (including Hamas, Palestinian Islamic Jihad [PIJ], and the al-Aqsa Martyrs Brigade) face a strong challenge from Israeli defensive measures, including surveillance assets and the barrier wall being constructed to prevent entry into Israel from the West Bank and Gaza. These groups have adopted a number of responses, including avoidance and camouflage, a variety of approaches to avoid the defensive wall, and new weapons that maintain their offensive capabilities.

- **Jemaah Islamiyah (JI) and affiliated groups.** In Southeast Asia, JI and its affiliated groups face varied defensive measures across the multiple countries in which they operate. These groups have adopted deception and forgery to maintain their ability to move from country to country and operational and technical ways to evade weapon detection technologies, and they have made other changes in target selection and operations to preserve their capabilities and operational freedom.

- **Liberation Tigers of Tamil Eelam (LTTE).** In Sri Lanka, LTTE used suicide terrorism for high-priority offensive missions. A number of defensive measures were put in place, including operative profiling, detection methods, and hardening potential targets of attack. LTTE responded by modifying its operational practices to include out-of-profile operatives, evading detection tech-

---

[3]   Although examining this was beyond the scope of this study, it should also be noted that the nature of the defensive technologies available and their application also shapes the "defender's perspective" about appropriate responses to the terrorist threat and assumptions about terrorist behavior.

niques or hiding the signatures they were designed to detect, and improving its techniques for penetrating defenses.

- **Provisional Irish Republican Army (PIRA).** In the United Kingdom, PIRA faced a diversity of defensive technologies aimed at undermining all facets of its operations. Through innovation and various operational approaches, PIRA developed strategies to counter security force information gathering and measures to jam or neutralize the group's weapons, protections around key targets, and even the ability of police to investigate and gather evidence after attacks.

Although the terrorist groups developed a wide variety of counter-technology measures for specific defensive technologies, many specific countermeasures they adopted have common elements that permit us to define a smaller number of fundamental countertechnology strategies. The groups applied these strategies, singly or in combination, when faced with a defensive technology threat. They are as follows:

- **Altering operational practices.** By changing the ways in which it carries out its activities or designs its operations, a terrorist group may blunt or eliminate the value of a defensive technology. Such changes frequently include efforts to hide from or otherwise undermine the technology's effect.
- **Making technological changes or substitutions.** By modifying its own technologies (e.g., weapons, communications, surveillance), acquiring new ones, or substituting new technologies for those currently in use, a terrorist group may gain the capacity to limit the impact of a technology on its activities.
- **Avoiding the defensive technology.** Rather than modifying how it acts to blunt the value of a defensive technology, a terrorist group may simply move its operations to an entirely different area to avoid it. Such displacement changes the distribution of terrorism, and, although this may constitute successful protection in the area in which the defensive technology is deployed, the ability to shift operations elsewhere limits the influence that the technology can have on the overall terrorist threat level.

- **Attacking the defensive technology.** If appropriate avenues are available, a terrorist group may seek to destroy or damage a defensive technology to remove it as a threat.

Although specific terrorist countertechnology efforts occasionally may fall into more than one of these classes, this taxonomy of strategies provides a systematic way to consider how terrorist organizations might respond to a newly deployed defensive measure.

Addressing Terrorist Countertechnology Efforts in Homeland Security Planning and Decisionmaking

The potential for terrorist groups to develop and deploy countermeasures for new defensive technologies must be addressed to ensure that protective efforts are effective and resources are allocated wisely.

Lessons for the Design of Defensive Technologies

To ensure that new defensive technology systems provide the greatest potential security benefits, they must be designed with terrorist countertechnology behaviors and past successes in mind. The efforts of the groups studied here suggest four techniques or approaches to use in developing plans for new defensive technologies.

- **Red teaming technology systems.** Given terrorist countertechnology behaviors, there is a clear need to test or "red team" new technologies, drawing on the terrorists' available palette of counterstrategies, to assess the limits of a technology before it is built and deployed.
- **Assessing adversary information requirements.** There is a clear need to analyze the information an adversary would need to circumvent the defensive technology and assess how the adversary might gain access to that information.
- **Designing flexibility into defensive technologies.** For most defensive measures, terrorist groups will eventually develop counterstrategies that limit their value. As a result, systems that are

flexible—that are not locked into specific modes of operation and can adapt themselves—may provide an added value.

- **Anticipating how technologies will guide terrorist adaptation.** When challenged by a new defensive technology, a successful terrorist effort to adapt may actually build it into a more potent threat than existed before the technology was deployed. To limit the potential for such unintended consequences, the design process for defensive systems should explore the effect of terrorist countertechnology responses not only on the value of the defensive systems, but also on the group overall and the nature of the threat it poses.

Lessons for Planning the Technological Components of Homeland Security Efforts

When terrorists are successful in countering all or part of the functioning of a defensive technology, the utility of the system may be significantly reduced or lost entirely. Such losses devalue the costs[4] society pays to design, produce, field, use, and maintain the technology.[5] As a result, potential countertechnology efforts need to be included in planning in three critical areas:

- **Include terrorist countertechnology efforts in programmatic and cost-benefit analyses of defensive systems.** In assessing a novel technology and its cost, the risk that its development and deployment might fail to deliver promised benefits is an established component of management planning. Like the competitive risk that another firm will develop a superior product, rendering a company's investments meaningless when both reach the

---

[4] The concept of costs includes not only financial and materiel costs but also auxiliary costs such as any reductions in privacy and civil liberties or costs paid in time or inconvenience by the public as a result of implementation of the security measures.

[5] For a nation as large and populous as the United States, these costs can be considerable. For example, at the time of this writing, major initiatives regarding border security and critical infrastructure protection are under consideration. Given the scope of both problems and the resources needed to implement solutions, considering how terrorists might act to counter protective measures that are put in place is clearly critical.

market, successful terrorist countertechnology efforts can similarly destroy the competitive advantage of a new defensive system. This *countertechnology risk* must be assessed and included as part of program management above and beyond the technological and other risks inherent in the effort itself.

- **Consider the relative costs of countering a technology and the cost of the technology itself.** The cost that a defensive technology can impose on a terrorist group—in effort and resources required to either withstand or counter its effects—is one measure of its value. If the cost is great enough, the technology's effect can be decisive. The cost that the nation should be willing to pay for a technology system must be related to its potential effect on its adversaries. When a technology can be countered with little investment on the part of the terrorist, the balance is in the terrorists' favor. This is particularly problematic when a group can access countertechnology strategies from other sources—for example, through technology transfer from other terrorist groups—that could significantly reduce or eliminate costs to the group.[6]

- **Address multistep countertechnology activities in assembling security technology portfolios.** Although this discussion focuses predominantly on single-step interactions between terrorist groups and defensive technologies—a single response by a group to a deployed technology—real conflicts are multistep contests. In consecutive iterations of measure and countermeasure competition, the potential exists for the terrorist to eventually overwhelm even the most adaptable defensive technology and reduce it to uselessness. If and when that occurs, new options will be needed. Given the potential for such "adaptive destruction" of individual security approaches, planning must consider defensive technologies as a portfolio, maintaining possibilities for alternative approaches in the event that currently effective technologies are neutralized.

---

[6] A companion publication produced during this research project, *Sharing the Dragon's Teeth: Terrorist Groups and the Exchange of New Technologies*, by Kim Cragin, Peter Chalk, Sara A. Daly, and Brian A. Jackson, addresses this topic in detail.

Conclusions

Although technologies can provide an edge in the effort to combat terrorism, that edge can be dulled by terrorist countertechnology efforts. An understanding of past terrorist efforts to counter defensive technologies underscores the complexity of designing new systems to protect society from the threat of these violent organizations. This analysis suggests that, in designing protective measures, it should not immediately be assumed that the newest and most advanced technologies—the highest wall, the most sensitive surveillance—will best protect society from terrorist attack. Drawing on common metaphors for defensive efforts, a fortress—relying on formidable but static defensive measures—is a limiting strategy. Once a wall is breached, the nation is open to attack. Depending on the adaptive capabilities of the adversary, a defensive model built of a variety of security measures that can be adjusted and redeployed as their vulnerable points are discovered provides a superior approach to addressing this portion of terrorist behavior. However, whatever combination of models and measures is chosen, it is only through fully exploring an adversary's countertechnology behaviors that vulnerabilities in a nation's defenses can be discovered and the best choices made to protect the nation from the threat of terrorism.

Acknowledgments

The authors would like to acknowledge the many individuals who provided key information on the terrorist groups discussed in this book. Due to the sensitivity of the topic, we cannot identify them by name, but we are grateful for their help. The project would not have been possible without them. The specific nature of our study, coupled with the obvious sensitivity of many of the technologies deployed against terrorist groups, made the observations and insights of individuals close to the conflicts with relevant expertise critical to our analysis and understanding of this part of terrorist behavior.

At the Department of Homeland Security, we would like to thank Robert Ross for valuable comments on a draft version of the manuscript. The peer reviewers of the document, Paul Stockton of the Naval Postgraduate School and Jack Riley of RAND, also provided beneficial input that significantly improved the final report.

Among our RAND colleagues, a number of people gave generously of their time and insights to improve the output of this effort. In particular, we would like to thank Tom LaTourrette for sharing his insights on antiterrorism technologies, Dave Frelinger and Martin Libicki for wide-ranging discussions on defensive approaches to terrorism and terrorist efforts to defeat them, Michael Wermuth and John Woodward for comments on drafts of the report, and John Baker for critical contributions to the project's analysis of JI. We would also like to thank Jolene Galegher of RAND's Research Communications Group for her assistance in preparing and polishing the project manuscript.

Abbreviations

AFP	Australian Federal Police
ASG	Abu Sayyaf Group
CCTV	closed-circuit television
DFLP	Democratic Front for the Liberation of Palestine
Hamas	Harakat Al-Muqawama Al-Islamia (The Islamic Resistance Movement)
IDF	Israel Defense Forces
JCLEC	Australia-Indonesia Jakarta Centre for Law Enforcement Cooperation
JI	Jemaah Islamiyah
MB	Muslim Brotherhood
PFLP	Popular Front for the Liberation of Palestine
PIJ	Palestinian Islamic Jihad
PLO	Palestinian Liberation Organization
PIRA	Provisional Irish Republican Army
LTTE	Liberation Tigers of Tamil Eelam
MILF	Moro Islamic Liberation Front

SIM	subscriber identity module (a component of a cellular telephone)
SLAF	Sri Lankan Armed Forces
TOSIS	Tigers' Organisation for Security Intelligence Service (a unit of LTTE)
TUF	Tamil United Front
TULF	Tamil United Liberation Front
UAV	unmanned aerial vehicles
WBGS	West Bank and Gaza Strip

CHAPTER ONE

Introduction

The level of threat posed by a terrorist group[1] is determined in large part by its ability to carry out the chain of activities needed to build its organizational capabilities and bring those capabilities to bear in violent action (Figure 1.1). In the effort to thwart terrorists' efforts and protect the U.S. homeland, the components of that activity chain are key targets for intervention. Such interventions attempt to prevent groups from recruiting members and collecting resources and to detect efforts to plan and stage operations. They also aim to defeat operations in progress through intervention or defensive measures and to identify and capture terrorists after an attack to prevent them from acting in the future.

As part of homeland security efforts, technology plays a key role within a larger, integrated strategy to target terrorist groups' activities and protect the public from the threat of terrorist violence.

Because of the variety of efforts to combat terrorism and tools that have been developed to pursue them, the analyst examining tech-

---

[1] Although some of the substate groups discussed in this book use tactics that are not purely terroristic in nature—for example, mixing traditional military operations against opposing security forces with terrorist bombings or assassinations—we use *terrorism* and *terrorist violence* as generic descriptors of the violent activities of substate groups.

In this book, we adopt the convention that *terrorism* is a tactic—the systematic and premeditated use, or threatened use, of violence by nonstate groups to further political or social objectives to coerce an audience larger than those directly affected. With terrorism defined as a tactic, it follows that individual organizations are not inherently *terrorist*. We use the terms *terrorist group* and *terrorist organization* as shorthand for "group that has chosen to use terrorism."

**Figure 1.1
The Terrorist Activity Chain**

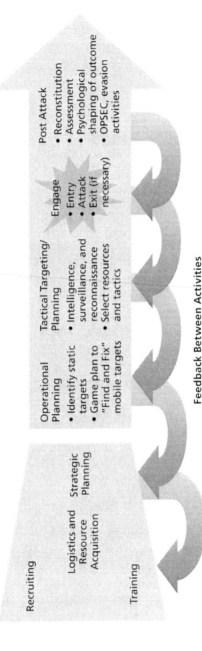

Capacity-Building and Planning Activities

Attack-Focused Activities

Recruiting

Logistics and Resource Acquisition

Strategic Planning

Training

Operational Planning
• Identify static targets
• Game plan to "Find and Fix" mobile targets

Tactical Targeting/ Planning
• Intelligence, surveillance, and reconnaissance
• Select resources and tactics

Engage
• Entry
• Attack
• Exit (if necessary)

Post Attack
• Reconstitution
• Assessment
• Psychological shaping of outcome
• OPSEC, evasion activities

Feedback Between Activities

NOTE: The RAND project team developed this model of terrorist activities, which is similar to other organizational activity models throughout the literature. The activity chain was used to provide a framework for analysis of the technologies in this book and to provide a common reference point for other technology-focused projects that were being carried out as part of this research effort, whose results are published separately.

RAND MG481-1.1

nological approaches in this area must consider a broad variety of technology types and, within each, many individual technologies designed to detect and frustrate terrorist efforts. Surveillance and other intelligence-gathering technologies aim to detect terrorist activities and provide law enforcement and other organizations with the information they need to dismantle terrorist cells. Direct countermeasures seek to disrupt attacks in progress, preventing groups from bringing their violent operations to fruition. Forensic science can be applied in the event that a group's operational plans are carried out, helping to lead investigators to the perpetrators and to support their arrest and prosecution. In all cases, technology systems seek to strengthen law enforcement and security organizations and enable them to protect the nation from the threat posed by extremist groups.

Defensive Technologies and the Effort to Combat Terrorism

Although many types of technology have roles to play in the overall effort to fight terrorism, this analysis focuses on a class of tools that we call *defensive technologies*—the systems and approaches deployed to protect an area and its citizens from terrorism by discovering and frustrating the plans of terrorists operating therein. These technologies were selected as the most relevant with respect to efforts to protect the U.S. homeland from the threat of terrorist attack.

As a result, a range of technologies applied in more proactive or offensive operations against terrorist groups—including military weaponry and similar technologies—are not included within the scope of this book. It should also be noted that the distinction between offensive and defensive technologies is admittedly ambiguous;[2] for example, the same intelligence-gathering system deployed in a defensive mode to

---

[2] This ambiguity similarly means that some activities we describe as defensive may be considered *counterterrorism* or *antiterrorism* depending on the specific definitions of those terms that are applied. For the purposes of this analysis, we therefore avoid use of that vocabulary and instead focus on the specific functions performed by technical systems in an effort to detect, prevent, protect from, and respond after a terrorist attack.

detect terrorist operations in progress could clearly gather information supporting offensive operations to combat terrorism. For technologies like these with dual applicability, we are interested in their deployment in a homeland security or defensive context.

The technologies that we have defined as defensive technologies can be organized into five primary classes that affect, in overlapping ways, sequential parts of the terrorist activity chain (Figure 1.2). The purposes of each of these types of technologies are as follows:

- **Information acquisition and management.** These tools include surveillance technologies and practices that enable law enforcement and security organizations to gather information on terrorist individuals, vehicles, and behaviors; to monitor sites and areas (including border information systems aimed at excluding terrorists from the country); to detect concealed weapons and operations in progress; and to maintain the profiles, databases, and systems needed to manage and use such information once collected.
- **Preventive action.** Technologies in this category include systems to counter specific terrorist weapon systems (e.g., radiodetonator jamming, antimissile systems) and systems designed to prevent terrorist access to money, weapons, technologies, and other resources or knowledge.
- **Denial.** Such approaches include traditional hardening of potential targets (e.g., setbacks, blast walls, reinforced windows, or other structures); design changes in potential targets to make them less susceptible to attack (e.g., increasing the robustness of infrastructure systems, immune buildings); hardening of the population (e.g., psychological preparedness efforts, vaccination); and security or guard force deployment.[3]
- **Response.** These technologies are designed to provide multiple capabilities, including defeating operations in progress (e.g., explosive ordnance disposal teams); ensuring that emergency

---

[3] Cameras and sensor systems deployed as part of efforts to protect targets—by monitoring for terrorist surveillance or attack—would fall into the first category of our taxonomy (acquiring and managing information) rather than being considered a protective or hardening measure.

Figure 1.2
Defensive Technologies Across the Terrorist Activity Chain

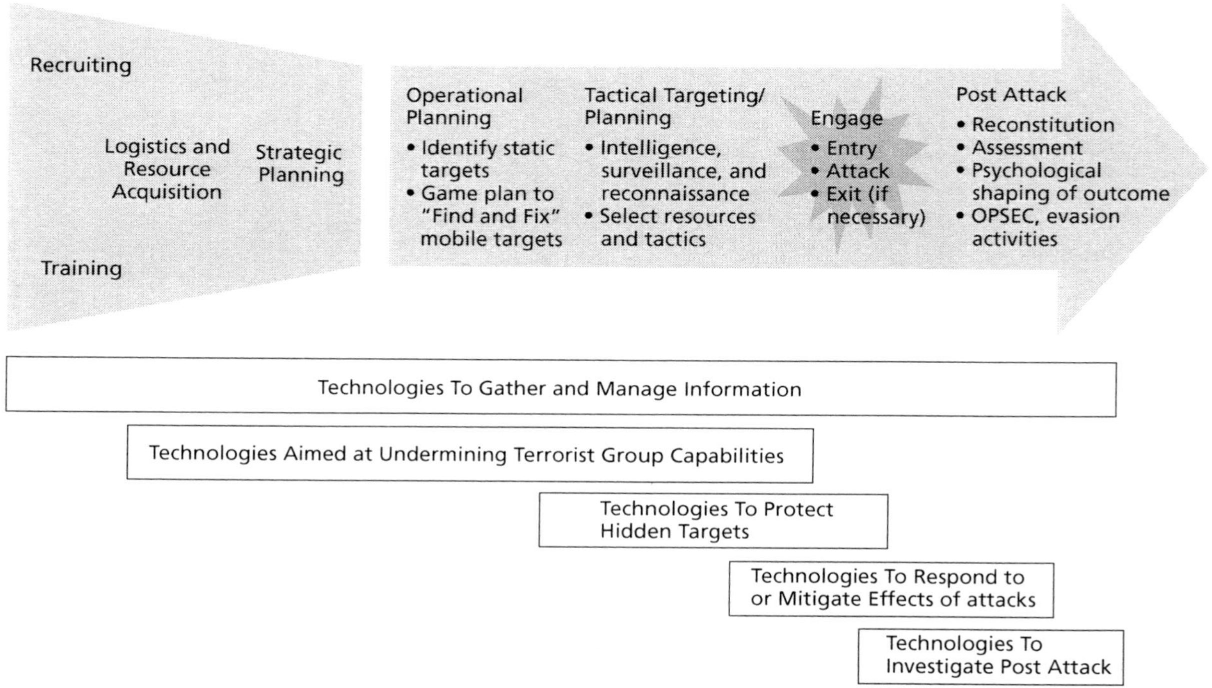

Recruiting

Logistics and Resource Acquisition

Strategic Planning

Training

Operational Planning
• Identify static targets
• Game plan to "Find and Fix" mobile targets

Tactical Targeting/ Planning
• Intelligence, surveillance, and reconnaissance
• Select resources and tactics

Engage
• Entry
• Attack
• Exit (if necessary)

Post Attack
• Reconstitution
• Assessment
• Psychological shaping of outcome
• OPSEC, evasion activities

Technologies To Gather and Manage Information

Technologies Aimed at Undermining Terrorist Group Capabilities

Technologies To Protect Hidden Targets

Technologies To Respond to or Mitigate Effects of attacks

Technologies To Investigate Post Attack

responses are adequate to treat casualties and limit the spread of damage from attacks in progress; coordinating response operations for increased effectiveness; making antidotes or other treatment methods for specific types of terrorist attacks readily available; and providing risk communication capabilities, which can be used to shape public responses to minimize the effects of an attack.

- **Investigation.** Technologies in this domain include forensic science and other investigative and identification technologies to analyze terrorist weapons, track and apprehend suspects, support prosecution of individuals responsible for terrorist operations, or enable other sovereign action against individuals or terrorist organizations.

This categorization, which reflects the wide variety of such technologies that have been developed, indicates that even limiting examination to the class of technologies we have defined as *defensive technologies* still captures a broad range of different technologies and systems.[4]

Although the contributions that technology can make in combating terrorism can be considerable, it should be noted that technology is only one in a range of strategies for combating terrorism. Law enforcement and security practitioners interviewed over the course of the study cautioned that the search for a technological "silver bullet" to address the problem of terrorism was unproductive and that, "if your security strategy relies only on technology, you are lost."[5] For example, virtually all sources consulted emphasized the preeminence of direct human intelligence—through infiltration of terrorist organizations or the recruitment of their members as agents—as the most important

---

[4] Capturing that diversity is important, because frequently a combination of defensive technologies or the combined outputs of multiple technology systems is required for successful homeland security efforts. At the same time, the diversity among these technologies can also hinder analysis by interfering with the ability to draw useful, crosscutting conclusions across technology categories. Breaking the overall class into a small number of subclasses for this work is intended as a middle path between these two extremes.

[5] Personal interviews with former law enforcement officials, England (May 2005) and with local officials, Indonesia, the Philippines, Singapore, and Thailand (March–April 2005).

element of an effort to combat terrorist activities.[6] Technologies—and defensive technologies in particular—were seen as a complement to those efforts. This book's emphasis on technical systems should not be interpreted as contradicting this view. Rather, the work described here should be seen as part of a multifaceted effort against terrorism to ensure that technology complements other efforts as effectively and efficiently as possible.

Terrorist Efforts to Overcome Defensive Technologies

Although the variety of defensive technologies available makes it possible to target many different components of the terrorists' activity chain, an effort to defend a nation against terrorism is not a one-sided game. Given the potential for defensive technologies to constrain the capabilities of terrorist groups and limit their operational freedom, these organizations are acutely aware of government efforts to deploy such countermeasures and actively seek ways to evade or counteract them. This measure-countermeasure, move-countermove dynamic is inherent in contests between organizations and, to the extent that the terrorists' efforts are successful, can significantly reduce or eliminate the value of defensive technologies.[7] This analysis examines efforts by terrorist groups in a variety of conflicts to neutralize defensive technologies.[8] Building on this historical assessment of terrorist activities, we

---

[6] Personal interviews with former law enforcement officials, England (May 2005) and with local officials, Indonesia, the Philippines, Singapore, and Thailand (March–April 2005).

[7] Although examining this was beyond the scope of this study, it should also be noted that the nature of the defensive technologies available and their application also shapes the "defender's perspective" about appropriate responses to the terrorist threat and assumptions about terrorist behavior.

[8] Just as we do not address security forces' efforts to infiltrate terrorist organizations in this book, we also do not examine terrorist organizations' efforts to infiltrate the security forces. Placing informers within police and intelligence organizations can provide terrorists with critical information and significant advantages in evading all types of defensive measures. Because our focus of analysis is the defensive technologies themselves and specific actions taken in response to their deployment, we did not examine terrorist infiltration activities as part of our analysis.

develop crosscutting conclusions aimed at improving the design and use of similar technologies in the current effort to protect the United States from terrorist attack.

About the Study

This research focuses on understanding terrorist group efforts to neutralize or defeat the utility of defensive technologies. To do so, the research team designed a method to draw relevant data from the history of a variety of terrorist conflicts and apply that information to broader technological questions relevant to current homeland security efforts and the ongoing effort to combat global terrorism. Our analysis involved a three-step process.

1. **Assess countertechnology behavior in specific terrorist conflicts.** To understand terrorists' efforts to counter defensive technologies deployed against them, the study team selected four individual terrorist conflicts for examination. To identify cases that were most relevant to current homeland security challenges, we chose to examine conflicts (1) in which national governments were engaged in substantial efforts to protect their home territories from the threat of terrorist attack, from groups operating within the nation or attacking from the outside; (2) in which governments had fielded a variety of defensive technological approaches against the groups, therefore providing a range of defensive technologies for our examination; and (3) in which the targeted terrorist organizations had survived and continued operations despite the government activities, therefore suggesting that their activities would provide as rich a data set as possible of terrorist countertechnology approaches. The following groups were chosen for examination:
    o Palestinian terrorist organizations, including Hamas (the Islamic Resistance Movement), Palestinian Islamic Jihad (PIJ), and the al-Aqsa Martyrs Brigade

o Jemaah Islamiyah (JI) and affiliated groups operating in Southeast Asia

o Liberation Tigers of Tamil Eelam (LTTE), particularly its suicide terrorism operations

o Provisional Irish Republican Army (PIRA)

Because case selection focused on conflicts that involved both a variety of defensive technologies and terrorist counter-technology efforts, a clear bias was produced in the range of terrorist behavior under examination. This approach inherently leads to a focus on comparatively sophisticated groups that face comparatively sophisticated law enforcement and security forces. This focus means that the behaviors we describe should not be interpreted as characteristic of all terrorist organizations. In the current environment, however, because the most significant current threat to the United States comes from comparatively sophisticated groups—al Qaeda and its ideological and operational affiliates—we believe that this bias in selection does not significantly limit the utility of the results of the analysis.

Examination of individual technologies and countertechnology responses was structured using a common set of defensive technologies (described in the section above),[9] organized into classes as shown in Figure 1.2. Because of the differences in these conflicts, the technologies fielded against each group differed. In addition, there are differences in the information available on the defensive measures and terrorist counterstrategies in each case. As a result, not all classes of defensive technologies discussed above are represented, or represented to identical degrees, in the chapters describing each terrorist group.

Individual researchers examined the data available in the open literature on the technologies that had been fielded against the groups and their efforts to counter them. The literature review

---

[9] Specific technologies within each class were identified based on the study team's previous experience, search of published literature and other information sources, and drawing on the outputs of other RAND homeland security technology research efforts. Examples of technologies that fall into each class are listed above.

was supplemented by examining other information sources and by interviewing experts in the academic, intelligence, and law enforcement communities who had direct experience with the groups being studied.

2. **Develop framework of terrorist responses to defensive technologies.** Supported by the information collected in the studies of individual terrorist groups, we developed a framework to capture the general methods that these organizations had pursued to counter defensive technologies.

3. **Assess implications of terrorist countertechnology behavior.** Building on the historical data, the research team examined how the terrorists had developed their counterstrategies, how broadly they could be applied to defensive technologies, whether they could be readily transferred to other terrorist organizations, and how important those behaviors are for the design of future defensive technologies.

Approaching the topic of terrorist countertechnology strategies from a historical perspective has significant strengths. Descriptions of actual terrorist attempts to circumvent technologies provide the most relevant data for assessing this behavior. Such a focus allows us to portray the ability of these groups to devise novel countermeasures, and, at the same time, takes into account the operational constraints that govern the technical activities of clandestine organizations.

Such an approach, however, also has clear limitations. Technologies relevant to contemporary homeland security efforts developed after the historical examples we examine will not be represented, and the examination will similarly be restricted to the technologies the selected groups faced and sought to counter. These limitations may mean that applying our findings will require extrapolation from analogous technologies that were fielded against the terrorist groups described.[10]

---

[10] In addition, a variety of broader questions regarding defensive technologies fall outside the scope of this analysis. This study examined defensive technologies, the terrorists' efforts to evade them, and how those efforts affect the value of the technologies. We did not examine, for example, important questions that have been raised in the United States and elsewhere about the effect of these technologies on society, especially with regard to their potential

Security issues also restrict the comprehensiveness of our analysis. As would be expected, much of the information on the technologies fielded against terrorist organizations and their efforts to penetrate them is classified. The research presented here is based on open-source materials and unclassified interviews with security experts and professionals; thus, our findings reflect only those technologies and terrorist counterstrategies for which open-source data are available. Although a similar classified review clearly would be valuable, we do not believe that the absence of such information significantly reduces the utility of the information presented here. Because of the broad variety of organizations—inside and outside government—involved in developing defensive technologies for homeland security, there is significant value in an analysis that can be shared that documents terrorists' efforts to defeat such systems. In addition, to the extent that there are similarities between the technologies described and relevant classified technologies, the results of this work should help to assess technologies that could not, for security reasons, be included in the analysis.

About This Monograph

This monograph synthesizes the results of the study, presenting lessons learned from past terrorist efforts to evade or circumvent defensive technologies. Chapters Two through Five describe the specific efforts of four terrorist organizations to counter a variety of defensive technologies deployed against them in their theaters of operation. Chapter Six assesses these past terrorist efforts, presents a framework describing strategies applied by terrorist organizations to counter defensive technologies, and describes crosscutting lessons drawn from our case studies. These lessons, we believe, are relevant to the design and deployment of future defensive technologies.

---

to constrain civil liberties. To the extent that these issues are raised at all, they are touched on as part of a broader discussion at the conclusion of the book on how terrorists' counter-technology efforts may affect consideration of or shift the cost-benefit analysis for particular defensive technologies—considerations and analyses in which the effect of technologies on society obviously must play an important part.

CHAPTER TWO

Palestinian Terrorist Groups

Introduction

This chapter explores how Palestinian terrorist groups, both secular nationalists and religious nationalists, have attempted to adapt and respond to Israeli counterterrorism technologies. First, it provides a brief background on political violence in Israel, the West Bank, and Gaza Strip (WBGS) as well as on the militant groups themselves. The chapter then discusses Israeli counterterrorism technologies and militants' responses, categorizing them as follows: (1) acquiring information about terrorist group members and their activities, (2) taking preventive action to undermine terrorist group capabilities, (3) denying terrorist access to targets through hardening measures, and (4) responding to terrorist attacks.

Importantly, Israeli security forces and Palestinian militants have engaged in periods of significant escalation and counterescalation over the past 50 years. As a result, changes in Israeli counterterrorism technologies sometimes are in response to new Palestinian tactics, rather than the other way around. We attempt to capture this dynamic as much as possible and provide some insight into the chronology and back-and-forth between these adversaries.

Political Violence in Israel, the West Bank, and Gaza Strip

Political violence is not new to Israel or WBGS. Even prior to the establishment of a Jewish state in 1948, British colonial authorities—under the British Mandate—struggled to contain terrorism and other forms of political violence conducted by Jews and Arabs alike. Having said that, Palestinian terrorism, directed against Israel and toward establishing a Palestinian state, arguably did not take the form of a concerted campaign until after the 1967 Six Day War. Israeli victory in the Six Day War signaled to the Palestinians and the rest of the Arab world in many ways that Israel could not be easily defeated, even by the combined efforts of Jordan, Egypt, and Syria. This realization also coincided with a rise in Palestinian nationalism, thus leading to a determination on the part of Palestinian nationalists to challenge the Israeli state on their own and not rely on supposed allies in the Arab world.[1] Yet, even though 1967 marked the beginning of Israeli occupation in the West Bank (formerly controlled by Jordan) and Gaza Strip (formerly controlled by Egypt), most Palestinian terrorism originated from outside WBGS from 1967 to 1998. Indeed, the Palestinian Liberation Organization (PLO), which functioned as an umbrella organization for multiple factions, operated out of Jordan, Lebanon, and Tunisia, not out of Israel or WBGS.[2]

The first *Intifada,* or literally "throwing off," significantly changed the dynamics of the conflict. In December 1987, an Israeli military vehicle crashed into a civilian truck, killing four Palestinians. Rumors quickly spread that the accident was deliberate, spawning a series of riots and protests throughout WBGS. The PLO and other Palestinian militant and political factions apparently were taken by surprise by these protests, as much as the Israeli government.[3] Yet these factions quickly mobilized under the Unified National Command of the Intifada. The Unified National Command's primary purpose was to organize the protests. Those responsible did this through the distribution of leaflets, from January 1988 through March 1989 (Meijer, 1998). Approximately 46 leaflets were distributed, although, according to Roel Meijer at the International Institute of Social History in Amsterdam, Shin Bet—an Israeli intelligence agency—published two

---

[1] This realization is perhaps best illustrated in Edward Said's writings at the time. For example, see Said (1979, p. xiii).

[2] For more information on the PLO, see Cobban (1984).

[3] For more information on the Intifada, see Mishal and Sela (2000) and Nassar and Heacock (1990).

of those 46 leaflets. Each leaflet had a distribution of between 35,000 and 100,000 copies (Meijer, 1998). The significance of the Intifada is that it did not, for the most part, employ violence. Stone throwing and tire burning were indeed part of the riots, but terrorism was not incorporated into the local protests. This lack of violence is significant, given that the Unified National Command was comprised of individuals from militant groups, including the Popular Front for the Liberation of Palestine (PFLP), the Democratic Front for the Liberation of Palestine (DFLP), al-Fatah (the militant wing of the PLO), and Hamas. Yet this strategy proved successful, as the international community pressured the Israeli government to engage in peace negotiations with the PLO, eventually leading to the 1993 Oslo Accords and subsequent Declaration of Principles.

If the period of the first Intifada lasted from 1987 to 1993, then the Oslo Period arguably lasted from 1993 to 1999. During this time, Palestinian militants still periodically engaged in terrorist attacks. Figure 2.1 illustrates the number of terrorist attacks per year in Israel and WBGS. Although these groups are described later, at this point, we should note that Hamas and PIJ rejected the Oslo Accords, and, thus, most of their attacks were aimed at disrupting this period of relative peace and quiet.

Figure 2.1
Terrorist Attacks in Israel, the West Bank, and Gaza Strip During the Oslo Period, 1993–1999

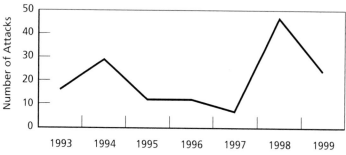

RAND *MG481-2.1*

In September 2000, then–Likud Party candidate Ariel Sharon visited East Jerusalem's most controversial site, known as the Temple Mount to Jews and the site of the Dome of the Rock and al-Aqsa Mosque to Muslims. To many Palestinians, this visit indicated Israeli intention to control this site and East Jerusalem, an area claimed by the Palestinian Authority and still under negotiation as part of the Oslo Accords. A series of riots broke out to protest this visit and the assumed statement, which eventually led to the al-Aqsa Intifada. Unlike the previous Intifada, this second round quickly escalated into violence. Figure 2.2 illustrates the nature of this violence. In comparison, the Palestinian Authority Web site states that 2,546 Palestinians have died as part of Israeli military incursions during the al-Aqsa Intifada (Palestinian Ministry of Foreign Affairs, undated). Significantly, unless otherwise noted, our book focuses on Israeli counterterrorism technologies and militant responses during the al-Aqsa Intifada.

The Geography of Israel, the West Bank, and Gaza Strip

The nature of the violence in Israel and WBGS is shaped, in part, by geography. According to the CIA World Factbook, approximately 1.4 million people live in the Gaza Strip, primarily concentrated in Gaza City and the Rafah, Khan Yunis, and Dayr al Balah refugee camps. The Gaza Strip is approximately 360 square kilometers, or twice the

Figure 2.2
Terrorist Attacks During the al-Aqsa Intifada, 2000–2005

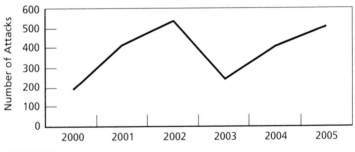

RAND MG481-2.2

size of Washington, D.C., but it has almost three times the population. Only one crossing exists between the Gaza Strip and Israel; a security fence surrounds the rest of Gaza's borders.

In comparison, approximately 2.4 million people live in the West Bank, which is 5,860 square kilometers. The West Bank has less than twice the population of Gaza, but over 15 times the space. The West Bank shares a 307-kilometer border with Israel and a 97-kilometer border with Jordan. As the al-Aqsa Intifada began to escalate, Israeli security forces could—because of the security fence—more easily control the flow of militants from Gaza into Israel, but they struggled with the borders between the West Bank and Israel. Thus, they began to build a security fence in the summer of 2002. We discuss this fence further as a counterterrorism technology below. Construction of this fence has been controversial, in part because it has altered the geography of the conflict. Significantly, the fence route is not dictated by the 1967 borders of the West Bank, but rather by security concerns. Thus, it tends to follow topography and geopolitical lines—taking the high ground in more rural areas and dividing Palestinian towns and Israeli settlements in more urban areas. This route has cut into territory traditionally considered Palestinian and part of the West Bank, annexing it to Israel proper.

Figure 2.3 illustrates the geographic distribution of terrorist attacks in Israel, the West Bank, and Gaza Strip since September 29, 2000. The largest dots indicate that more than 30 attacks have taken place in these areas: Jerusalem, Tel Aviv, and the border crossing between Israel and Gaza. Despite the new security fence in the West Bank, Jerusalem still represents a key point of crossing between Palestinian and Israeli territories; plus it remains a significant symbolic target. It therefore is logical that a number of attacks would occur in this city. Likewise, Tel Aviv remains an important target due to its being a political and population center. The presence of settlements along the border between Gaza and Israel as well as the security precautions by Israeli security forces both explain the large number of attacks along this border. Medium-sized dots illustrate that fewer than 30 attacks have occurred, and the smallest dots indicate fewer than five attacks. Our data come from the RAND-MIPT Terrorism Incident

Figure 2.3
Distribution of Terrorist Attacks Since September 2000

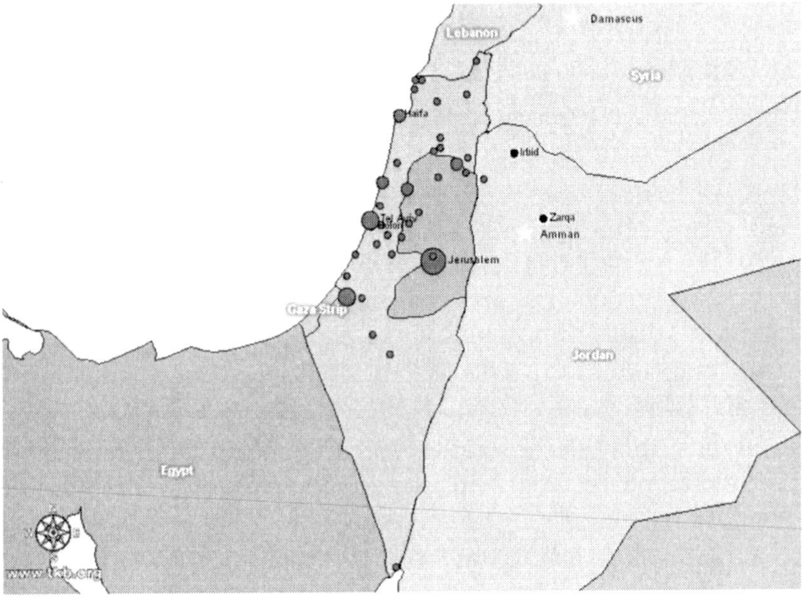

RAND *MG481-2.3*

Database (National Memorial Institute for the Prevention of Terrorism, undated).

Secular-Nationalist Militants:
al-Fatah, PFLP, Tanzim, Force-17, and al-Aqsa Martyrs Brigade

Today, Palestinian militants can be divided into two basic categories, secular nationalists and religious nationalists. The five most active groups of secular nationalists are al-Fatah, the PFLP, Tanzim, Force-17, and al-Aqsa Martyrs Brigade. This section provides a brief overview of these militant groups.

Fatah, a reverse acronym for Harakat al-Tahir al-Filastiniyya, has dominated the PLO since the late 1960s (Cobban, 1984). Led by Yasser Arafat until his death in December 2004, most Fatah leaders joined the Palestinian Authority during the Oslo Period. Because Fatah had sanctioned peace negotiations between Israel and the Palestinian Authority, it could not officially participate in the al-Aqsa Intifada. Thus, other associated militant groups took on this role. However, many former leaders of Fatah had a hand in the violence of the al-Aqsa Intifada.

Perhaps the best example is Marwan Barghouti and the Tanzim. Barghouti reportedly joined Fatah's Ramallah branch in the mid-1970s at the age of 15.[4] He played a leadership role during the first Intifada and, in some ways, became the personification of those who remained in Palestine, while many Fatah and PLO members—for example, Arafat—fought the Israelis from abroad. During the al-Aqsa Intifada, Barghouti criticized Arafat and other Palestinian Authority members for rampant corruption. But he remained a key player in the Fatah-Tanzim group. Israeli security forces arrested Barghouti in April 2002 (Cobban, 1984). After Barghouti's arrest, the Tanzim's role in the al-Aqsa Intifada dissipated.

Like the Tanzim, Force-17 is the name given to Arafat's personal security service. In March 2001, an Israel Defense Forces (IDF) spokesperson accused Force-17 of engaging in terrorist attacks. These attacks included three drive-by shootings that killed eight and wounded 22. The spokesperson argued that Mahmud Damarah, Force-17's Ramallah leader, had also provided weapons to other Palestinian militant groups in the area.[5] Like the Tanzim, Force-17 does not appear to have been a significant player in the al-Aqsa Intifada after 2002.

Another militant group linked to Fatah is the al-Aqsa Martyrs Brigade. Unlike the previous two groups, this militant group emerged out of the al-Aqsa Intifada. Its loosely linked cells have been responsible for a significant amount of the violence in Israel, the West Bank, and Gaza. Although sometimes the rhetoric from al-Aqsa has religious overtones, it is strictly a secular-nationalist group. Attacks attributed to this group include the following:

- In January 2002, members of the al-Aqsa Martyrs Brigade opened fire on a bat mitzvah party in Hadera, killing six and wounding 35.

---

[4] For more information, see BBC News (undated).

[5] For more information on technology exchanges and relationships between different Palestinian militant groups, see Cragin et al. (forthcoming).

- In March 2002, a member of the al-Aqsa Martyrs Brigade conducted a suicide bombing in Jerusalem's ultraorthodox neighborhood, Me'a Sha'arim, killing nine and wounding 45.
- In November 2002, members of the al-Aqsa Martyrs Brigade attacked a Likud Party headquarters in Beit She'an, killing six and wounding 43.
- In July 2003, members of the al-Aqsa Martyrs Brigade in Gaza fired mortar shells on a settlement, injuring no one.
- In January 2004, a member of the al-Aqsa Martyrs Brigade conducted a suicide bombing on a bus in Jerusalem, killing eight and wounding approximately 60.
- In September 2004, members of the al-Aqsa Martyrs Brigade fired two mortars on a settlement in Gaza, injuring no one.
- In January 2005, al-Aqsa Martyrs Brigade and Hamas both claimed responsibility for a suicide truck bombing at the Karni Crossing between Israel and the Gaza Strip. The attack killed six and wounded 15.

Notably, of the secular nationalists, al-Aqsa Martyrs Brigade is the only militant group to have adopted suicide terrorism. In fact, this group surpassed Hamas in the number of suicide bombings and casualties during the al-Aqsa Intifada.

Finally, the PFLP has also been somewhat active in the al-Aqsa Intifada (see PBS Frontline, 2002).

Religious-Nationalist Militants: Hamas and Palestinian Islamic Jihad

In addition to the secular nationalists described briefly above, two religious-nationalist militant groups also have operated in Israel during the al-Aqsa Intifada. We use the term *religious nationalists* because, although these groups have religious objectives, these objectives are interpreted best in the context of the overall nationalist objectives of the Palestinian movement. In this sense, they represent a different type of religious militant group from al Qaeda.

Both Hamas and PIJ have roots in the Palestinian chapters of the Muslim Brotherhood (MB).[6] The primary goal of MB, in brief, is a religious revival. At its origins, Egypt's MB was a charitable and missionary organization. In the late 1930s, MB began to take on a more political nature, especially as it began to provide support to the Palestinians (Mitchell, 1969). Yet it continued to adhere to a platform of nonviolence. In 1979, students dissatisfied with MB's nonviolent approach formed PIJ in the Gaza Strip (Moghadam, 2003). At the time, an ongoing debate divided MB in Egypt into two factions: Supporters of the traditional nonviolent approach were opposed by a new generation of leaders who advocated for a top-down violent revolution. PIJ fell into the latter group.

In contrast, Hamas did not officially emerge out of MB until January 1988. Although Hamas decided to take a violent approach vis-à-vis Israel, it is strongly opposed to internecine violence in the Palestinian communities.[7] In this context, it still adheres to the MB objective of a nonviolent, grassroots, religious revival among the Palestinians. Thus, Hamas developed an expansive charitable network in WBGS, while PIJ concentrated on terrorist attacks. This ideological difference perhaps explains why Hamas continues to receive between approximately 5 percent and 25 percent approval ratings among Palestinians and PIJ does not.[8] Attacks by PIJ and Hamas during the al-Aqsa Intifada include the following:

- In June 2001, a suicide bomber detonated outside the Dolphinarium in Tel Aviv, killing 17 and injuring approximately 120 people. Although PIJ originally claimed responsibility, arrests eventually demonstrated that Hamas members conducted the attack.

---

[6] For more information on PIJ and Hamas' ideological roots, see Abu'Amr (1994).

[7] This opposition can be seen in its statements as well as its actions. See, for example, Zahhar (1995). In November 1994, Palestinian rioters attempted to tear down a Palestinian Authority prison and break out Hamas prisoners. Hamas officials were instrumental in calming the violence (see Harub, 2000).

[8] These figures come from the Centre for Palestine Research and Studies in Nablus, as well as the Jerusalem Media and Communications Center, which have charted public support for Hamas on a biannual basis since 1993.

- In October 2001, a remote-detonated device (car bomb) exploded in the Talpiot neighborhood of Jerusalem, with no casualties. PIJ claimed responsibility.
- In May 2003, a suicide bomber detonated in the French Hill neighborhood of Jerusalem, killing seven and injuring approximately 20 people. The bomber was disguised as a religious Jew.
- In March 2004, two suicide bombers detonated in Ashdod, killing 10 people. Al-Aqsa Martyrs Brigade and Hamas claimed that it was a joint attack.
- In July 2005, a Qassam rocket was fired at the Sederot settlement near the Gaza Strip. No one was injured, although the rocket damaged several cars and a house patio. PIJ claimed responsibility.

Now that we have provided some background on the militant groups involved in the al-Aqsa Intifada, the following sections outline Israeli counterterrorism technologies and how these groups have attempted to adapt to them.

Information Acquisition and Management

With regard to information acquisition, human intelligence represents the core of Israeli counterterrorism policy. Prior to the Oslo Period, the Israeli security apparatus maintained an expansive presence in WBGS, and this physical colocation provided it with the opportunity to recruit Palestinian informants. This recruitment system allowed the security services to stay one step—or several steps—ahead of the militants (Blanche, 2004; Alon Ben-David, 2004d). With regard to the importance of human intelligence, a former chief of the Israeli security service, Yakob Perry, recently stated,

> Modern Intelligence has wiretapping networks capable in picking up any telephone or radio conversation in the world; their code-cracking capabilities of their computer systems are virtually endless. And yet, all of that has not prevented the 11 September strike on the USA. This very fact has only confirmed a lesson that I learnt from decades of security-intelligence work: There is

no substitute for a human source who can supply advance alert of indication, and there probably never will be. Technology is an important, even vital element, but there is no substitute for people. (Eshel, 2002b)

After the Oslo Period, the Israeli military withdrew from many areas in the West Bank and Gaza. As its physical presence declined, some experts have argued that the quality of the Israeli government's intelligence also declined (Katz, 1999). Following the advent of the al-Aqsa Intifada, however, Operation Defensive Shield (April 2002) allowed the Israelis to reestablish a strong ground presence in WBGS. In Operation Defensive Shield, the Israeli military reentered the West Bank: "division-sized forces, including reserve brigades, were used in one of the largest 'cordon-and-search' operations ever mounted by the Israel Defense Force" (Eshel, 2002a).[9] It was followed by Operation Determined Path, which saw special commando teams conducting systematic house-to-house searches (Eshel, 2002a). Although these operations could logically be viewed as preventive action or denial, it is important to emphasize that they provided the Israeli government with the opportunity to gather significant intelligence—for example, through the confiscation of planning documents or membership lists and the detention of militants—and reestablish a presence critical to the success of its human intelligence network.

Technologies Deployed

Although human intelligence is the core of information acquisition for the Israeli security community, it incorporates technology as well. For example, the Israeli military has deployed a number of static and mobile surveillance cameras. These technologies allow the security authorities to monitor the flow of suspicious individuals into areas with either

---

[9] Such operations as those described here have multiple goals beyond gathering information on terrorist groups, including undermining terrorist capabilities as described below. "The three week operation's goal was to attack the infrastructure of Palestinian terrorism. The IDF hoped to catch as many terrorists as possible, to discover and destroy arms caches and bomb-making laboratories, and to gather the necessary intelligence to thwart future attacks" (Catignani, 2005, p. 256).

high-probability targets or likely gathering sites for potential terrorists. The surveillance cameras often work in conjunction with a series of security checkpoints—discussed below under "Denial"—especially in times of high alert.

The Israeli police also reportedly have incorporated mobile intelligence systems into security vehicles to aid in the search for individuals once human intelligence provides warning of an impending attack.[10] This capability has proven important in the Israeli response to potential suicide attacks. For example, in September 2001, the first Israeli Arab suicide bomber killed three people at a railway station in Naharia (National Memorial Institute for the Prevention of Terrorism, undated). Police reportedly had intelligence on this individual, including his name and likeness, for four days prior to the attack but still could not intercept him.[11] This mobile intelligence system allows the security police to disseminate information quickly and respond to such threats more rapidly.

In addition, the Israeli government is known to have broad communication-interception capabilities. Reportedly, the Israeli government has specialized units within various security and intelligence services (such as the military intelligence unit referred to as "8200") focused on signal intelligence in fighting Palestinian militants [Alon Ben-David, 2004c]). The Israeli government has used AH-64 Apache helicopters as well as unmanned aerial vehicles (UAVs) to monitor electronic communications in the West Bank and Gaza Strip (WBGS) as thoroughly as possible (Alon Ben-David, 2004d, 2005; *Jane's Sentinel Security Assessment*, 2005). The Israeli government uses airborne technologies to overcome UHF and FM radio degradation, which occurs frequently in high-density urban environments (Eshel, 2002b). In effect, these signal intelligence technologies add another layer of security to the ground surveillance and human intelligence systems.

---

[10] Author interview with Israeli police authorities, Haifa (June 2003).

[11] Author interviews with Israeli police authorities, Haifa (June 2003).

Countertechnology Responses by Palestinian Groups

Palestinian militants have taken a number of steps to mitigate the impact of these counterterrorism policies and technologies. First, with regard to informants, the Palestinian community has taken a strong stance against "collaborators" since the 1948 war. Collaborators range from individuals who might sell their property inside the West Bank and Gaza to Israelis, enabling an encroachment into Palestinian territories to informants for the security services (Palestinian Academic Society for the Study of International Affairs [PASSIA], 2001). The latter, in particular, face the possibility of assassination by Palestinian militants or even mobs. For example, *The Guardian*, a UK-based paper, reported that one accused collaborator was shot and hung from an electricity pylon in Hebron in 2002 (McGreal, 2004). Hamas reportedly has a special unit, named *Jehaz Aman*, which investigates potential collaborators as well as certain new members (Katz, 1999). According to some sources, more than 1,000 individuals were killed as collaborators during the first Intifada (PASSIA, 2001).

Second, militants attempt to avoid Israeli signal intelligence technologies. For example, cell phones are given to friends or cousins to organize missions. Callers are instructed to keep conversations to a minimum and to use code words.[12] Additionally, militants change their phones, limiting calls as much as possible immediately prior to operations. Hamas and PIJ leaders have taken specific actions to further diminish members' cell phone use for attack coordination by prohibiting using cell phones during operations (Eshel, 2002b). These groups also use faxes, couriers, coded leaflets, and the Internet to transfer instructions between cells (Eshel, 2002b). It appears that some al-Aqsa Martyrs Brigade cells requested help from Hizballah to improve their communication security. For example, in March 2004, Israeli authorities arrested a leader of the al-Aqsa Martyrs Brigade's Khan Yunis cell in Gaza. He reportedly stated that a Hizballah representative came

---

[12] Personal interview with Israeli police officer, Tel Aviv (May 2005), with Israeli scholar (May 2005), and with IDF representative (May 2005).

to Gaza in 2003 to improve their information security practices and technology.[13]

Third, Palestinian militants also have attempted to reduce the effectiveness of aerial surveillance technologies. To do this, militants have covered alleyways and streets with sheets and carpets (O'Sullivan and Abu Toameh, 2004).[14] This approach is more likely to be effective in urban environments, such as refugee camps in the Gaza Strip. In rural areas, children have been known to watch for helicopters and UAVs from rooftops in the WBGS (Moore, 2004). Militants also limit the duration of their mortar attacks to two minutes so they can escape before Israeli helicopters arrive (Fighel, 2005a).

Finally, in an attempt to counter mobile and static surveillance cameras, Palestinian militants continue to explore different disguises: For example, terrorists have dressed as Israeli soldiers and even religious Jews.

Preventive Action

Preventive action, along with human intelligence, is a cornerstone of Israeli counterterrorism policies and technologies. At times, this preventive—or preemptive—approach has been controversial in the international community, especially with regard to assassinating bomb-makers and key militant leaders. Notably, with the advent of the al-Aqsa Intifada, Israeli security authorities have shifted to focusing on denial technologies more and more. In part, this shift could be the result of recognition that, during periods of high escalation, some militants are likely to get through this preventive layer.

Technologies Deployed

As implied above, a key aspect of Israeli preventive action is the assassination of bomb-makers. To conduct these assassinations, Israeli secu-

---

[13] For more information on this particular incident, see Cragin et al. (forthcoming) and "ISA Arrests Head of Gaza Strip Hezbollah Cell" (2004).

[14] Personal interview with IDF representative (May 2005).

rity forces have used booby-trapped mobile phones and vehicles, snipers, helicopters or F-16s, and even poison. For example, in January 1996, Israeli security authorities assassinated Yahya Ayyash, Hamas' "Engineer." At the time, Israeli authorities made the argument that they only assassinated ticking bombs, not political leaders. But this policy changed during the al-Aqsa Intifada. For example, Israeli authorities assassinated Sheikh Yasin, Hamas' key political leader, in March 2004. So it appears that the Israeli security services broadened their list of targets to include spiritual and political leaders, in addition to bomb-makers and operational planners.

Israel also has transferred lessons learned against Hizballah and its remote-detonated devices in southern Lebanon to the WBGS.[15] These technologies include the use of jamming technologies to prevent remote detonation, especially for secondary devices (Eshel, 2002b).

Beyond these two main preventive technologies, the Israeli security services have adopted several others, including the following:

- An antimissile system gives Israeli residents a warning (approximately 15 to 20 seconds) prior to mortar attacks. The system uses loud speakers to give residents time to seek cover.
- An EGIS system detects explosives at checkpoints. Israeli authorities use this system at the Rafah checkpoint in Gaza.
- The illegal weapon market has been "salted" with dysfunctional bullets or detonation devices.

Finally, Israeli authorities have turned to technology in an attempt to prevent Palestinian weapon smuggling. Palestinian militants have built a series of tunnels under the "Philadelphi corridor," which denotes a narrow border—approximately 4 kilometers long and 100 meters wide—between Rafah and Egypt to smuggle weapons, people, and goods into the Gaza Strip. We discuss these tunnels below under "Denial," because they are the Palestinians' key response to Israel's security fences. With regard to preventive action, however, IDF researchers reportedly are exploring sensors to detect underground

---

[15] For more information, see Cragin (2005).

excavation (Susser, 2005). Prototypes have allowed IDF engineers to detect digging or tunneling sounds; a specialist unit is then deployed to further determine the tunnel's exact location (Frisch, 2005; Grinberg, 2005).

Countertechnology Responses by Palestinian Groups

As with information acquisition, Palestinian militants have taken a number of steps to reduce the effectiveness of Israeli preventive action. In most responses, militants have adopted new modus operandi, rather than exploring new technologies.

First, to avoid assassinations, targeted militants avoid using personal vehicles for transportation. They live in crowded urban centers and travel surrounded by supporters to increase the risk of civilian casualties for the Israeli government. So, for example, with the aforementioned assassination of Hamas leader, Sheikh Yasin, Israeli authorities also killed seven other Palestinians, four of whom were civilians, and wounded an additional 15 people (Anderson and Moore, 2004). By surrounding potential targets with civilians, militants raise the cost of the Israeli assassination policy.

Second, in responding to jamming technologies, militants allegedly have increased surveillance of potential targets to identify those most vulnerable or perhaps those without explosive-detection technologies (Dudkevitch, 2002). Palestinian militants also have modified explosive devices used for suicide bombings, developing devices that can be hidden in jacket linings or with smaller, more easily concealed belts (Lefkovits, 2002).

Finally, militants have adjusted their targeting to include areas outside security entrances, first responders, and IDF checkpoints. For example, in December 2001, two suicide bombers detonated explosive belts at Ben Yehuda Mall in Jerusalem. A secondary device (remotely detonated or timed-device car bomb) detonated when responders arrived at the scene. Hamas claimed responsibility for this attack. This tactical adjustment corresponds with a shift by the militants to adopt a more guerrilla-warfare–like approach, attacking IDF convoys and military targets inside the WBGS, since they have experienced some difficulties getting inside Israel proper.

Denial

Since the advent of the al-Aqsa Intifada, Israeli authorities have begun to adopt a variety of denial technologies, layering them on each other as well as on top of human intelligence and preventive measures. As with the preventive measures discussed above, the militants' counter-responses appear to be more tactical adjustments than the adoption of new technologies.

Technologies Deployed

With an increase in the level of violence, Israeli authorities quickly set up checkpoints along the main roads between Israel and WBGS. This response is not unusual, but rather is a typical Israeli countermeasure. For example, after a series of suicide bombings by Hamas in early 1996, Israeli authorities established checkpoints to monitor the flow of vehicles in and out of Israel. In the past, these checkpoints served a small security purpose, but their primary objective was to put pressure on the Palestinian economy and force public support against the militants. Now the checkpoints' purpose is fundamentally security.

Israelis have implemented both fixed and random mobile checkpoints. The random searches normally are implemented after intelligence forewarns of an imminent attack or after an attack in an effort to catch accomplices. These checkpoints provide an opportunity for Israeli authorities to screen travelers as well as to complicate, and perhaps deter, the movement of Palestinian attackers. Passage through these checkpoints requires identification and frequently an identity check against a centralized registry (IDF, 2004). New technology also allows Israelis to verify Palestinian ID cards with biometric devices that scan hands and faces (Copans, 2003). Known as the Basel Project, this combination of "smart cards" and biometric scanning is designed to allow for efficient and effective border crossings through the security checkpoints (Morgenstern, 2003). Virtually all Israeli officials and scholars interviewed for this book indicated that the extensive layering of checkpoints greatly contributes to their ability to slow the flow of suicide bombers, bomb-making materials, and weapons in and out of WBGS.

In addition to these military and police checkpoints, Israeli authorities began to build a barrier around the West Bank approximately one year into the al-Aqsa Intifada. Authorities determined that the fence surrounding the Gaza Strip was a key factor in the limited number of terrorists who had infiltrated into Israel from that area. Thus, they reasoned, a fence would similarly work if it surrounded the West Bank. A map of the route of the security barrier is shown in Figure 2.4.

Notably, the barrier has two main components. First, in areas where Palestinian and Israeli towns are colocated, concrete walls provide a solid protection against gunfire. These walls are similar to the one that separates Loyalist and Republican territories in Belfast. Second, in more remote areas along the barrier route, a security fence prevents infiltration by Palestinian militants into Israel. This fence is chain-linked with barbed wire on top, along with electronic sensors and cameras (O'Sullivan, 2003). Space has been cleared on either side of the fence, so that security vehicles can travel quickly from guard posts to any potential area of penetration. Additionally, the fence is built on top of a foundation of concrete. Thus, the technology is designed to prevent individuals from climbing over or digging under this barrier.[16]

Countertechnology Responses by Palestinian Groups

Palestinian militants have tried a number of different tactics to counter Israeli checkpoints. The primary tactic—to avoid them—has become increasingly difficult with the addition of the security fence discussed above. Perhaps this explains the attacks that take place at checkpoints along the border between Israel and WBGS. For example, in April 2002, militants killed two people and wounded seven when they threw grenades at a checkpoint near the Eretz crossing in northern Gaza Strip. PIJ claimed responsibility for this attack. Although it has become difficult for Palestinians to get through these checkpoints, some militants have managed to penetrate into Israel. For example, some groups have used female suicide bombers, hoping that cultural sensitivities might make it easier for them to avoid security ("Female Bomber a Hamas

---

[16] Author interviews and visit to the security fence in Israel (June 2002).

Figure 2.4
Route Map of Israeli Security Barrier

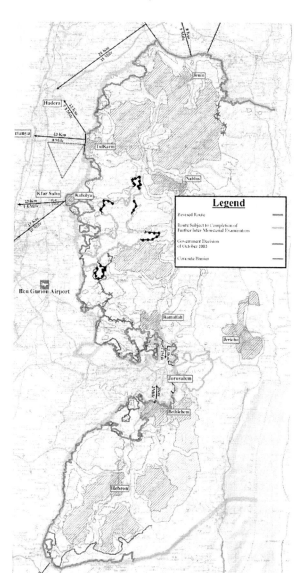

SOURCE: Excerpted from Israel Ministry of Defense (2005).
RAND *MG481-2.4*

First," 2004; Beyler, 2003). As mentioned in previous sections, other militants have dressed in IDF uniforms also in an effort to avoid being stopped (Reeves, 2001). These adjustments on the part of Palestinian militants have caused most Israeli authorities to determine that "profiles" of suicide bombers are not useful in denying attacks.

Similarly, Palestinians have gone beyond the use of female bombers and disguises to avoid the security checkpoints. *Haaretz* reports, for example, that militants in the West Bank have cooperated with car thieves to circumvent the as-yet incomplete wall (Harel, 2005b). Vehicle license tags are color-coded to help Israeli soldiers determine what vehicles have been registered in Israel (yellow), the West Bank (green), and Gaza Strip (blue). By obtaining stolen vehicles with yellow license tags, militants apparently hope that they can more easily pass through checkpoints.

Perhaps the most significant of Palestinian militants' technological responses is the development and use of Qassam rockets. Militants use these rockets to target settlements, such as Sederot, or Israeli cities, such as Ashkelon, which are just across security perimeters (Fishman, 2004b, 2005b, 2005c; Blanche, 2003; Richardson, 2002). *Jane's Missiles and Rockets* (Richardson, 2002) reports that the weapons are

> manufactured from easily available materials that require only a simple, short manufacturing process. The raw materials used include water pipes (used in the manufacture of the rocket motor casing, some parts of the warhead and some of the rocket tail structure), sheet steel and aluminum (to produce the gas exhaust nozzle, warhead and fuse), and potassium nitrate fertilizer and powdered sugar (for the manufacture of the propellant). All of these materials are available in Palestinian areas throughout Judea, Samaria and the Gaza Strip.

Militants have improved upon the range of the Qassam rocket, increasing its reach from 9 to 12 to 14 kilometers during the al-Aqsa Intifada (Richardson, 2002, 2004; Alon Ben-David, 2003). According to some authors, militants have also deployed Qassam rockets with increasingly large warheads (Alon Ben-David, 2005). Militants conducted approximately 94 rocket attacks in 2003, but over three times more (308) one year later in 2004 (National Memorial Institute for the Prevention of Terrorism, undated).

Having said that, the Qassam rockets are notoriously inaccurate. Their one advantage is the ability to reach over security barriers. As a result, militants have produced and deployed Qassam rockets only in the Gaza area. But as Israeli authorities complete the barrier around the West Bank, and, if the peace process falters once again, militants could attempt to use them in the West Bank. In fact, some reports indicate that Hamas has already made a concerted effort to extend such operations to the West Bank. But five attempts at large-scale production of Qassam rockets there have been foiled (Fighel, 2005b). Some fear that the increased use of rockets may also stimulate more intense efforts to acquire surface-to-air missile technology, which would enable terrorist groups to challenge Israel's air superiority (Harel, 2004b).[17]

Finally, other techniques have been used to circumvent the security barrier in the West Bank. In particular, some militants reportedly have deployed specially crafted ladders that enable them to climb over the security fences without detection by the sensors at the top (Elon, 2002). Others identified a key vulnerability—water drains without security grates—though Israeli authorities have fixed that oversight.[18]

Palestinian organizations have also engaged in tunneling activity to circumvent barriers, predominantly to support weapon smuggling from Egypt to Gaza (Fishman, 2005b; Blanche, 2004; IDF, 2003). The development of tunnel smuggling networks is a result of the difficulty of transporting weapons and explosives into and out of the West Bank and Gaza. The IDF notes that Palestinians have taken a number of measures to avoid detection of their tunneling operations, including building tunnels in residential areas (entrances are often through private homes and property), digging at night, transporting displaced dirt and sand, and staging diversionary strikes against IDF outposts

---

[17] Additional efforts to harden potential targets have been taken in response to rocket and mortar attacks. Israel began hardening the roofs of houses, schools, and other buildings near the Israel/Gaza border in June 2005 (Harel, 2005b). As yet, there is no clear Hamas or PIJ response to these efforts, though both groups continue to work on improving the rockets' distance and accuracy.

[18] Author interviews with Israeli engineer for the security fence (June 2002).

to conceal the sound of explosives. In addition, tunnel entrances are often hidden behind false walls or under showers and sewer lids (IDF, 2003). Recently, there has also been evidence of tunneling under portions of the security fence, as well as tunneling to get beneath IDF posts. Groups have reportedly also incorporated their own countermeasures into the tunneling efforts, such as planting booby traps in tunnels dug beneath sensors at the security fence (Blanche, 2004; Fishman, 2004c).

Response

Israeli authorities respond quickly to terrorist attacks, in part to reduce media attention for the militants and in part to observe burial rules in Jewish law. But the aforementioned secondary device at Ben Yehuda Square also demonstrated the potential danger that first responders face in Israel. So Israeli authorities developed the C-Guard EXP System, which can jam cellular phones within a 1-kilometer operational radius (see Netline, undated). This technology can then help first responders prevent a secondary explosion, detonated remotely by a cellular phone. Because this system is approximately the size of a briefcase, bomb disposal teams can deploy it easily. At this point, Palestinian militants have not developed an alternative technology to cellular remote-detonation devices.

Conclusion

Israeli security authorities and Palestinian militants have been engaged in an armed struggle for approximately 50 years, during which both have adapted their tactics and technologies to challenge the opponent. This chapter, however, focuses primarily on the al-Aqsa Intifada. Countertechnology responses by Palestinian groups are summarized in Table 2.1.

Table 2.1
Palestinian Groups' Technological Innovations:
Purpose and Intended Mitigation of Government Countermeasures

Innovation	Purpose	Intended Mitigation of Government Countermeasures
Changing cellular phones and maintaining communication discipline	Maintain operational security	Signals intelligence collection
Prohibiting cell phone use during operations	Maintain operational security	Signals intelligence collection
Covering alleyways with sheets and carpets in urban environments	Enable covert movement	Aerial surveillance
Using watchers to identify and alert members to aerial vehicles	Limit surveillance effectiveness	Aerial surveillance
Reducing duration of attacks	Ensure escape of operatives	Rapid aerial response capabilities
Using disguise	Avoid surveillance efforts; penetrate security measures	Variety of Israeli aerial and ground surveillance capabilities; security barrier
Modifying transportation practices and use of civilian shields	Protect known operatives from attack	Israeli rapid response and strike capabilities
Increasing reconnaissance of potential targets to identify sites uncovered by countermeasures	Increase chances of operational success	Explosives detection and jamming technologies
Reducing the size and increasing concealability of weapons	Increase chances of operational success	Surveillance and security efforts aimed at detecting operations in progress
Substituting softer, less protected targets such as first responders or exposed sites (e.g., checkpoints)	Avoid security measures to increase chances of operational success	Security and hardening measures
Using operatives who break profile (e.g., women)	Penetrate security measures	Security and hardening measures; security barrier

Table 2.1—Continued

Innovation	Purpose	Intended Mitigation of Government Countermeasures
Cooperating with smuggling organizations to circumvent security	Penetrate security measures	Security barrier
Tunneling	Penetrate security measures	International border checkpoints; security barrier
Using stolen vehicles with license tags designed to avoid suspicion	Penetrate security measures	Security barrier
Using alternative weapons (Qassam rockets) that allow attack over the security barrier	Stage attacks despite security measures in place	Security barrier
Using ladders to scale security barrier	Penetrate security measures	Security barrier

In addition, our research yielded the following broader conclusions.

New technologies adopted by Israel served to constrain terrorist operations in the short term. The combination of extensive surveillance, physical barriers, and preventive action appears to have reduced the number and lethality of attacks by Palestinian militants in Israel, the West Bank, and Gaza Strip. For example, the fact that militants have begun to attack checkpoints and military targets within WBGS indicates that they have had difficulty penetrating into Israel. Similarly, the surveillance technologies appear to have limited militants' ability to communicate and coordinate their operations. Finally, the C-Guard EXP system is likely responsible for limiting the number of secondary devices that the militants have been able to detonate successfully.

Palestinian militants mostly responded to Israeli technological superiority by adjusting their tactics. All of the militant groups in our study responded to new counterterrorism technologies by adapting their tactics. For example, al-Aqsa Martyrs Brigade conducted an attack using a militant dressed as a religious Jew. Hamas similarly forbade its members from using cellular phones during operations. These and other groups also learned to limit their mortar attacks to two minutes so that they could evade Israeli response. These examples demonstrate that terrorists do not necessarily have to engage in a "technology war" to counter government responses or even escalate the conflict. Tactical changes can be sufficient.

Sometimes, Palestinian militants sought out new technologies to respond to Israeli counterterrorism measures. In some instances, Palestinian militants have sought out new technologies. These technologies were not sophisticated per se but allowed the militants to add a new weapon to their arsenal. The most significant example is the development of the Qassam rockets, which have allowed militants to overcome the security barrier, albeit to a limited degree. Another example is the remotely detonated car bomb that was used as a secondary device in the Ben Yehuda Square attack. Although not a new technology, militants' use of carpets to cover walkways in the Gaza refugee camps and defeat Israeli aerial surveillance is perhaps one of the more interesting countermeasures. It demonstrates that even a low-level response can confound some of the most sophisticated technologies.

In sum, the Israeli experience indicates that new technologies alone will not make a decisive difference in a struggle against terrorism. By limiting operational effectiveness, however, they can reduce the short-term threats that terrorist groups pose.

CHAPTER THREE

Jemaah Islamiyah and Affiliated Groups

Introduction

Since 2000, JI has undertaken terrorist activities in an effort to establish an Islamic caliphate extending from southern Thailand, through the Malay Peninsula, across the Indonesian archipelago, and into the southern Philippines (International Crisis Group, 2002). Founded in Malaysia in 1995 by Abu Bakr Baasyir and Abdullah Sungkar—radicals in hiding from the Indonesian government—JI relocated to Indonesia in the aftermath of dictator General Suharto's fall from power in 1998.

By 1998, JI had allied itself with al Qaeda (Arabic for "the base"). JI was never formally subordinated to al Qaeda. Some analysts view JI as subordinate in practice, but others note that JI has generally prioritized regional objectives over al Qaeda's global objectives (Abuza, 2004; Baker, 2005; Ressa, 2003; National Commission on Terrorist Attacks upon the United States, 2004). In December 2001, Singaporean authorities arrested several JI members who had been involved in preparations with al Qaeda operatives for undertaking attacks, some of them planned suicide attacks, against four foreign embassies, Western business interests, U.S. Navy ships, an airbase, and a major water pipeline between Singapore and Malaysia (Baker, 2005).

JI has cooperated with many insurgent groups in the region, especially the more fundamentalist Islamic insurgents, such as the Moro Islamic Liberation Front (MILF), especially in the nearest and most porous areas of the southern Philippines and Malaysia. The differentiation between JI and other terrorist groups is often ambiguous. As with al Qaeda, JI's influence on local terrorism can be notional or inspirational rather than material. While JI may contribute money, training, expertise, or even leadership to local terrorist activities, the true value of JI's contribution remains unclear. Terrorists often claim or appear to be members of multiple terrorist organizations, and they can draw tactical support, such as the use of safe houses, from non-JI members who share their Islamic separatist aspirations. Meanwhile, individual motivations can appear confusing, such as when money-making ventures may fund personal lifestyles rather than the group's terrorist activities. JI appears to act alone mostly in Indonesia and operates outside Indonesia mostly in cooperation with indigenous groups. For instance, JI bombed Manila in December 30, 2003, in cooperation with factions from the MILF, which was formed to fight for an autonomous Islamic state in the southern Philippines (Tan, 2004).[1] Some of these groups operate largely as insurgencies, while JI has been more focused on conducting terrorist attacks.[2]

The majority of JI's terrorist activities in Southeast Asia[3] to date have been directed against symbolic Western targets. Recent incidents for which responsibility has been claimed by, or reliably attributed to, JI include the following:

- Jakarta Australian Embassy bombing, September 2004, nine dead

---

[1] For an introduction to the MILF insurgents and related jihadist activity, see Tan (2004).

[2] Insurgencies, historically, have been the main source of domestic terrorism in the region, although most insurgent activities are not best described as terrorist. Instead, most insurgent activities are bomb attacks on security convoys, armed ambushes of security patrols, shootings of individual soldiers or local political leaders, or organized criminal activities for profit. Most of the insurgents are best described as secessionists with ethnic or ethnoreligious claims to secession. For instance, the government of the Philippines has been fighting an insurgency in the southern Philippines, mainly on the island of Mindanao, associated with Muslim Filipinos, some of Indonesian descent. Similarly, the Thai government is fighting an insurgency in southern Thailand associated with Muslim Thais of Malay ethnicity.

[3] Southeast Asia is normally considered to include the countries of Brunei, Burma, Cambodia, Laos, and Vietnam. JI is less active in these latter countries but has operated in all of them.

- attack on the J. W. Marriott Hotel in Jakarta, August 2003, 12 dead
- Bali nightclub attacks, October 2002, more than 200 dead
- series of explosions at churches in Jakarta, Sumatra, Lombok, Java, and Batam Island, December 2000, 15 dead
- attempted assassination of the Filipino ambassador to Indonesia, August 2000, three dead.

JI is also believed responsible for funding bombings of the metro in Manila on December 30, 2000, in which 27 people died.

JI's organizational structure is geographically based and hierarchy-driven. Operational responsibility is divided into four territories, called *mantiqis,* the leaders of which comprise the group's central decisionmaking body:

- *mantiqi 1:* Malaysia, Singapore, Southern Thailand, and Cambodia
- *mantiqi 2:* Indonesia
- *mantiqi 3:* Borneo and Southern Philippines
- *mantiqi 4:* Australia.

Recruits to populate the territories' branches, platoons, and squads are solicited from radical Islamic boarding schools and are trained in the network of JI camps in the southern Philippines. JI's membership is currently estimated to exceed 1,000, with several hundred believed to be operationally oriented (Jones, 2003; Globalsecurity.org, undated).

Despite the death of Sungkar in 1999 and the capture and detention of over 300 JI members and key operatives in the years since 2001, most experts believe that JI retains the capacity and will to launch attacks throughout Southeast Asia and describe the group as being highly committed, resourceful, flexible, and adaptive (Vaughn et al., 2005). By analyzing JI responses to Indonesian, Philippine, Australian, and Singaporean deployments of antiterror defensive technologies, this chapter examines the extent to which these characterizations are true. Specifically, the sections that follow examine JI reactions to the deployment of four distinct classes of defensive countermeasures:

- **information acquisition and management:** technologies intended to gather and manage information or restrict JI movement
- **preventive action:** technologies intended to degrade JI's operational, logistical, or planning capabilities
- **denial:** technologies intended to harden targets against attack
- **investigation:** technologies intended to produce successful investigations after an attack occurs.

Each section is comprised of a description of the specific countermeasure deployed and examples of JI responses. Discussion of the implications of this dynamic for future selection and allocation of defensive resources concludes the chapter.

Information Acquisition and Management

Acquisition of information about the intentions, capabilities, and activities of terrorist organizations is of paramount importance to states seeking to defend themselves against attack.

Technologies Deployed

Given the importance of intelligence data, states frequently deploy multiple technologies—from the cultivation of human sources to the use of earth-orbiting satellites—in an effort to gather intelligence. The variety of such information acquisition technologies deployed in Southeast Asia today is both a function of extant resource disparities among states and a reflection of the challenges inherent in defending against the dispersed nature of the terrorist organizations with which the region must contend.

Overhead surveillance technologies. Overhead surveillance technologies are those used to locate and monitor visually detectable terrorist group activities, such as the movement of persons or equipment at training camps or weapon development facilities. Systems in use by countries around the world today include sophisticated earth observation satellites capable of providing high-resolution images of targeted geographies, UAVs equipped with advanced cameras and radar systems, and traditional piloted aircraft.

In Southeast Asia, indigenous military forces have made extensive use of piloted overflights to survey territory believed or known to be occupied by JI and other groups. With foreign forces having deployed UAVs in the region, many of the Southeast Asian officials interviewed for this study expressed interest in acquiring similar capabilities; budget limitations, however, have thus far prevented any significant movement in this direction.[4] Recently, some countries in the region have benefited from satellite images shared by foreign countries or those purchased from commercial enterprises.

Communication monitoring. By the end of 2000, most indigenous intelligence agencies in Southeast Asia had acquired technologies for intercepting cell phone conversations. The successful application of these technologies, however, has proved difficult, with officials noting that the targeting of specific individuals remains dependent upon prior acquisition of good human intelligence.[5]

According to a Filipino intelligence offer, the recent arrest of a prominent terrorist demonstrates the technology's limitations. In that case, a female informer supplied the suspected terrorist's cell phone number to authorities. Officers received permission to intercept the suspect's phone calls and proceeded to tap the line, receiving audio access to his conversations but no information about his location ("Azahari's Blinking Cell Phone," 2003).[6] Indeed, the suspect was arrested only after the informer invited him to a shopping mall, where the suspect was visually identified upon answering the informant's phone call.[7]

---

[4] Personal interviews with local officials, Indonesia, the Philippines, Singapore, and Thailand (March–April 2005).

[5] Personal interviews with local officials, Indonesia, the Philippines, Singapore, and Thailand (March–April 2005).

[6] Another good example of this problem was the ability of Indonesian police to monitor the apparent movements of key JI bombers, Azahari Husin and Noordin Mohammad Top, in the months before and after their bombing attack of the J. W. Marriott Hotel in Jakarta in August 2003 without being able to pinpoint their location sufficiently to track them down. See "Azahari's Blinking Cell Phone" (2003).

[7] Personal interview with intelligence official, the Philippines (March 2005).

Closed-circuit television (CCTV). South Asian intelligence and law enforcement agencies are increasingly turning to the use of CCTV technology to detect and monitor individuals acting suspiciously near potential targets.[8] Most densely distributed in Singapore, CCTV cameras were initially deployed in an effort to deter prostitution and petty crime. Recently, however, police throughout the region have noted the important contributions CCTV can make to terror investigations, referring specifically to an incident in which CCTV cameras filmed a suspect carrying a backpack near the blast site at the Hat Yai airport in Thailand moments before the explosions of April 3, 2005 ("Two Die in Triple Hat Yai Blasts," 2005; "Blasts Hit Airport, Hotel and Carrefour," 2005). Subsequent to these events, Thai security and intelligence chiefs agreed to install CCTV cameras in 40 "safety spots" throughout Bangkok, with the number intended to increase to 88 by May 2005 ("Tighter Security at Airports," 2005; "Country on Security Alert," 2005, pp. 1, 4).

Border security. Border security has been, and remains, uneven across the countries of Southeast Asia. Philippine and Indonesian coastlines are expansive and difficult to patrol and, as late as 2004, were widely considered to be highly porous. Indeed, both countries have acknowledged that the length and complexity of their maritime borders and limited naval resources greatly hinder the implementation of adequate security controls (Ramakrishna, 2004).

Indonesia, further, maintained lax visa requirements for travelers from Muslim states throughout 2000–2004; Malaysia has only required such visitors to acquire visas at all since 2002 (Abuza, 2004; Ressa, 2003; National Commission on Terrorist Attacks upon the United States, 2004). None of these countries, further, has yet computerized its immigration information, continuing instead to rely on manual records.[9]

---

[8] Personal interviews with local officials, Indonesia, the Philippines, Singapore, and Thailand (March–April 2005).

[9] Personal field research, Indonesia, the Philippines, Singapore, and Thailand (March–April 2005).

Singapore, by contrast, currently deploys modern technologies and requires visitors from most countries to have visas. It uses a computerized database to track individuals crossing its borders and in 2004 deployed Sentinel, a technology designed to detect unauthorized changes to passports (Ramakrishna, 2004). Iris and fingerprint recognition technology enable paperless border crossings for frequent travelers, an automated screening system reads and reports license plate numbers, and radiographic scanners survey the interiors of vehicles for illegal cargo.[10]

These measures have led to significant successes for Singaporean counterterrorism authorities. In 2000, detained JI members provided details of Manila bomber Fathur Rahman al-Ghozi's aliases and approximate dates of entry into Singapore. A subsequent database search of entry and exit cards provided details of his passport, information that was then provided to Filipino authorities and led to his arrest in a Manila hotel room in January 2002.[11]

Countertechnology Responses by JI and Its Affiliates

Terrorist organizations can pursue a number of techniques to neutralize the effectiveness of discovered or assumed information acquisition technologies. Prominent among these are efforts to avoid detection and identification by using false documents, frequent movement between and among geographic areas, pursuit of illegal activities only outside monitored areas, or modification of behavior within those areas so as to avoid arousing suspicion. Groups may also seek to prevent disruption of operations by using individuals whose characteristics are incon-

---

[10] Personal interviews with public officials, Singapore (April 2005). In the future, Sentinel may take on a facial recognition technology, which will match the passenger's features against the passport photograph. Singapore will be one of the few countries to issue biometric passports before the fall 2005 deadline declared by the United States. A chassis scanner is under evaluation; it would be installed in the top surface of roads to scan the underside of a vehicle's chassis to determine whether the vehicle has been adapted to carry illegal cargo, such as explosives.

[11] Personal interview with public official, Singapore (April 2005). Al-Ghozi later escaped from jail and was killed on October 12, 2003, during a confrontation with Filipino security personnel.

sistent with those of common or expected terrorist profiles. JI has used each of these tactics in response to the region's deployment of information acquisition technologies.

Overhead surveillance. JI has sought to neutralize the effectiveness of overhead surveillance by relocating, reducing, and camouflaging its base and training activities. During the 1990s, JI successfully established training camps in Indonesia, Malaysia, and the Philippines, and perhaps in Australia as well (Abuza, 2003b; Almonte, 2004; Gunaratna, 2004; Rabasa, 2003). The most notable of these, Camp Abu Bakar, was located on the island of Mindanao and destroyed by the Philippine government in July 2000. Since that time, JI has leveraged contacts with Philippine terrorist organization MILF to acquire access to its network of more than 20 training camps. Obscured from aerial surveillance by jungle and mountainous terrain, JI's access to these remote and protected geographies allows the group to circumvent higher-risk areas in Indonesia, Malaysia, and Singapore (Ressa, 2003).

Communication monitoring. In response to communication monitoring systems, JI has adapted its practices over time to reduce the risk of being detected, understood, and located by security forces. Seemingly aware of U.S. successes in compromising the security of satellite phones, the group uses them sparingly in remote areas but not elsewhere.[12] In urban areas, JI continues to rely upon cell phones for communication, but, rather than using voice-based applications, the group is increasingly using the technology's text-messaging capability, exchanges law enforcement agencies find more difficult to trace (Ressa, 2003). Where person-to-person contact is necessary, calls are short and cryptic. Bali bomber Imam Samudra, for example, limited his calls to 20 seconds, and JI members consistently use code words in conversation (Turnbull, 2003).[13]

Subscriber identity module (SIM) cards—the chips that identify each phone and owner—are, further, either changed frequently or purchased prepaid from suppliers that rarely require buyer identification

---

[12] Personal interview with defense official, the Philippines (March 2005).

[13] Personal interviews with public officials, the Philippines (March 2005) and Singapore (April 2005).

information (Tortermvasana, 2005).[14] Where such recourse is unavailable, JI members have paid unwitting local collaborators to purchase and register cell phones on their behalf; the locals may never know they have assisted a terrorist, except in those rare instances in which the phone itself is captured by authorities (Ressa, 2003).

JI members routinely rely on couriers to deliver verbal messages and are also known to have used email to communicate. There is little evidence, however, to indicate use of more sophisticated forms of Internet communication (Gunaratna, 2002; Ressa, 2003).[15] Although Imam Samudra, likely the most technologically competent member of JI, allegedly embedded messages in digital images, the skill is not believed to be widespread, and Samudra himself is now on death row, convicted of having coordinated the Bali bombings of 2002 (Sipress, 2004).

Border control. JI has pursued a number of techniques to overcome border security technologies deployed throughout Southeast Asia. Confronting increased scrutiny at airports, for example, JI members now appear to avoid air travel, instead seeking out obscure crossings and favoring boats, buses, or trains. "Freelance terrorist" Khalid Shaikh Mohammed, who worked with JI before his capture in Pakistan in 2003, in fact advised one of the Bali bombers to use buses and trains rather than planes because security was tightest at airports (Ressa, 2003; National Commission on Terrorist Attacks upon the United States, 2004). Maritime travel in particular has proved an attractive option, with JI members successfully emigrating from Indonesia to the southern Philippines and Malaysia via undeclared coastal landings.[16]

JI has also increased its acquisition and use of aliases and false passports, and there are indications that the group is willing to discard such documentation after only one use.[17] Indeed, JI seems to have regarded Bangkok, Thailand, as an ideal logistical base and transitional

---

[14] Personal interviews with public officials, the Philippines (March 2005) and Singapore (April 2005).

[15] Personal interview with intelligence official, the Philippines (March 2005).

[16] Personal interview with defense official, the Philippines (March 2005).

[17] Personal interview with public official, the Philippines (March 2005).

node in part because false documentation is readily available in the city and easy to acquire (Abuza, 2004; Ramakrishna, 2004; Singh, 2004).

JI members traveling across borders, further, are likely to fabricate their nationalities to reduce the chances of being noticed and are increasingly of unexpected or atypical profiles.[18] Women have been used to purchase and transport materials, for example, while converts to Islam willing to use their non-Muslim names and extraregional passports have been recruited to carry out operations.[19]

CCTV. In an effort to avoid detection by CCTV cameras, JI has begun to attempt to disguise its public activities, most particularly its surveillance of potential targets for attack. Members of one of JI's Singaporean cells pretended to be joggers while they were, in fact, photographing water pipelines at a nature reserve. A JI member casing a mass transit station in Singapore brought his children along to give the impression that his intentions and activities were benign (Ressa, 2003; Singapore Ministry of Home Affairs, 2003).

Preventive Action

Technologies capable of attacking and degrading the resources and tools available to terrorist organizations can be of great use in diminishing the threat they pose.

Technologies Deployed

In their effort to undermine JI and its affiliated groups, the countries of Southeast Asia have incorporated many of these technologies into their security regimes, including the deployment of weapon detection, detonation prevention, and financial tracking devices.

Weapon detection. Weapon detection technologies include those used to find weapons hidden on the person, in luggage, or in cargo. Singaporean detection capabilities, the most advanced in the region,

---

[18] Personal interview with local officials, the Philippines (March 2005) and Singapore (April 2005).

[19] Personal interview with defense official, the Philippines (March 2005).

include three-dimensional scanning technologies that are particularly useful for locating weapons hidden in luggage. Other nations' systems remain fairly basic and are comprised of the use of sniffer dogs, proscription of certain items on public transportation, and physical inspection of passengers and baggage.[20] In Manila, for example, local transit authorities responded to reports that some terrorists were concealing explosive devices in sardine cans by banning passengers from carrying tin cans inside subway trains and stations ("JI Militants Said Plotting 'Major' Attacks on US," 2005).

Detonation prevention. Technologies intended to prevent terrorists from exploding their devices are also available and in use in Southeast Asia. JI has adopted the use of cell phone detonators to such an extent that, for most officials in the region, evidence of the technique's use is considered an almost conclusive indicator of JI involvement.[21] In response, most security authorities in the region have acquired cell phone jammers, mostly with foreign assistance.[22] These jammers interrupt signals over a limited range, but, because they disrupt normal cell phone activity, they are usually deployed only in response to very good intelligence or in the immediate aftermath of a bombing.[23]

Financial tracking. Security organizations have used technology to monitor terrorist access to bank accounts with considerable success. In November 2002, Indonesian police pinpointed the location of one of the Bali bombers—Imam Samudra—when he withdrew cash from

---

[20] Personal interview with border official, Singapore (April 2005).

[21] Personal interviews with public officials, Indonesia, the Philippines, Singapore, and Thailand (March–April 2005).

[22] For instance, Thailand acquired cell phone jammers in November 2004 (Davis, 2005b).

[23] Before the proliferation of cell phone detonators, clocks or watches, first mechanical and then digital, often were used as detonators. Unlike timers, cell phones permit remote-controlled detonations, as do two-way radios and remote door bells. The bombers either use the cell phone's alarm function or place a call to the cell phone. Cell phones are usually used to place the call, both because a mobile observer can place the call and because cell phone connections are often more reliable than are land lines in the region. Additionally, cell phones acquired in one country may be used to detonate bombs in another country. For instance, most bombs now detonated in Thailand are apparently detonated by cell phones acquired in Malaysia, where they are cheaper and more difficult to track.

an automated teller machine in Banten. Information provided by an accomplice and Samudra's own family seems, however, to have pointed authorities in the correct direction originally, making the intelligence gained from the financial transaction monitoring less important (Sebastian, 2004; Turnbull, 2003). Security officials have also used technology to mine data on financial transfers in order to spot terrorist fund transfers.

Countertechnology Responses by JI and Its Affiliates

JI has responded aggressively to attacks on its resources and preferred attack techniques, countering state technologies by adjusting weapon transport methods, implementing redundancies in its detonation systems, and capitalizing on noninstitutional means for transferring funds.

Weapon detection. In general, weapon detection equipment seems to have encouraged JI to disaggregate its weapons and transport the pieces separately; to carry explosives in multiple, smaller containers; and to hide bombs in innocuous-seeming packages or cargos. In March 2005, Philippine authorities discovered that a local terrorist organization had collected 1,300 pounds of explosives. After further investigation, intelligence officials concluded that half of the cache had been transported in packages weighing no more than a few pounds— including in everyday items such as toothpaste tubes and cookie tins— and had reason to believe that JI had recommended the technique to the group.[24]

Intelligence officials also believe that JI terrorists are experimenting with mechanisms for masking identifying vapors by packing explosives within food stuffs or mixing powders with pungent items such as coffee or cardamom.[25] Philippine authorities know that smugglers have hidden contraband under rotten fish and believe that JI engages these same smugglers or uses their techniques (Bell, Larson, and Haynes, 2004–2005). In an apparently successful effort to conceal the explosives

---

[24] Personal interviews with intelligence officials, the Philippines (March 2005).

[25] Personal interviews with intelligence and defense officials, the Philippines and Thailand (March 2005) and Indonesia (April 2005).

in the vehicle used in bombing the J. W. Marriott Hotel on August 5, 2003, JI bomb designer Azahari Husin reportedly concealed the bomb in the rear of the van under piles of plywood to avoid arousing the suspicions of security guards (Wijayanta, 2003).

Detonation prevention. Definitive or specific data on JI's response to the use of cell phone jammers are not readily available. It is known, however, that other terrorist organizations have succeeded in defeating the technology by detonating bombs in quick succession in one area or by targeting widely separated locations, and it is considered likely that JI will, or already does, mimic these tactics. JI is also known to rely upon redundant means, such as timers or suicide bombers, for detonating explosives, a practice that, although perhaps not employed specifically to defeat jammers, would certainly do so.

Financial tracking. JI has consistently pursued avenues of asset management and funds transfer that considerably degrade the effectiveness of financial tracking technologies. In the 1990s, the group established a network of international religious schools, front companies, and Islamic charities—from which it diverts legitimate and illegitimate donations alike—to raise and move money. JI accesses these funds through *hawalas,* unregulated banking systems based on trust, in which money is made available internationally without a physical transfer ever taking place or a transaction record being generated (Abuza, 2003a, 2004).

When regular banking channels are used, most flows are routed through the globally connected and comparatively weakly regulated institutions of Malaysia, Thailand, and the Philippines. JI also has used Cambodian banks, but with less frequency given their circumscribed international reach (Abuza, 2003a, 2004; Ramakrishna, 2004; Turnbull, 2003). In an effort to curtail terrorist access to large sums of money, some regional governments have reduced the size of allowable transfers. Not surprisingly, rather than discouraging such transactions in general, these policies have only encouraged groups to undertake multiple transfers of smaller quantities, driving authorities to lower thresholds even further. The Filipino Congress's ban of transfers in excess of $80,000, for example, was subsequently revised to prohibit those surpassing only $10,000 (Abuza, 2003b).

JI has also made tentative steps toward engaging in financial fraud to generate revenue. Convicted bomber Imam Samudra used proceeds from a robbery and online credit card scheme to fund the 2002 Bali attacks and, despite being in detention, published advice on hacking and financial fraud (Sipress, 2004). Despite this encouragement and guidance, however, JI may not have much need to engage in these comparatively high-risk activities: Malaysian intelligence officials believe that the man responsible for JI's Malaysian and Singaporean operations had $500,000 in assets at his disposal for use in operations, a sum greater than the estimated cost of the 9/11 attacks (Abuza, 2003a).

Denial

Denial is understood here to refer to the deployment of technologies intended to prevent attackers from reaching their target. The most prevalent form of denial technology is target hardening.

Technologies Deployed

A target is considered "soft" if its location, structure, or function render it vulnerable to attack or make it difficult to protect. Such targets can be "hardened" against attack with the deployment of a combination of passive and active measures. Passive hardening techniques include defensive site design, security walls, and protective fences. Active measures, such as checkpoint screening of individuals, vehicles, and cargo, or increased police presence, are often deployed to bolster a site's more static means of protection.[26]

Southeast Asian countries and foreign government and commercial industries with interests in the region have hardened their assets considerably over the last four years, with embassies now surrounded by barriers and hotels and tourist spots heavily patrolled by security officers. These measures have met with some success—the presence of guards and controlled traffic flow around the J. W. Marriott Hotel in

---

[26] The use of cameras as an element of hardening potential targets was addressed previously in the section on information gathering and management.

Jakarta, for example, prevented JI's large vehicle bomb from reaching the hotel lobby, limiting the number of casualties the attack was able to produce (Ressa, 2003).

Countertechnology Responses by JI and Its Affiliates

The hardening of potential targets in its operational area has elicited three responses from JI: The group has adjusted its operations to increase the likelihood of reaching the target, enhanced the yield of its bombs, and demonstrated increased interest in soft targets. In October 2001, Fathur Rohman al-Ghozi, a senior JI operative and bomb-making expert, convinced JI leadership to abandon plans to attack the U.S. embassy in Manila, noting it was set back from the road and well protected by barriers and security guards. For the Bali and Jakarta attacks—both on soft targets—bombs composed of a combination of explosive materials meant to boost their destructiveness were delivered by individuals willing to die during the operation, increasing both the likelihood the bombs would reach their target and the lethality of their detonation in the event that unexpected impediments were encountered along the way (Ressa, 2003; Wijayanta, 2003). The Bali operation was also designed to ensure successful explosion of the largest vehicle bomb, with multiple redundancies in place—a cell phone detonator, a manual switch, a timer, and a trigger rigged to release if the bomb package were opened (Ressa, 2003).

Target hardening has not always successfully discouraged or defended against JI operations. Improved security at the Australian embassy in Jakarta—including reinforcement of the outside wall and installation of shatterproof windows—did not prevent the deaths of nine people during a JI attack in September 2004 (Sipress and Nakashima, 2004). This suggests that, if JI feels compelled to attack a target, perhaps because of its symbolic or material value, it may not be deterred by target hardening.[27]

---

[27] JI may have preferred softer targets, such as hotels, which received extra security after the United States identified hotels as potential targets a week before the attack (McBeth, 2001). On the other hand, JI may have been weakened by earlier arrests of experienced members, without whom the attack was less sophisticated than earlier JI attacks (Jones, 2004).

Investigation

The capacity to successfully investigate attacks and apprehend perpetrators is fundamental to a state's, or region's, ability to deter and undermine terrorist activity.

Technologies Deployed

Proper investigative procedure, adequate preservation of crime scenes, and the use of the full array of available forensic technologies are, accordingly, competencies the countries of Southeast Asia are seeking to improve. Domestic authorities have begun to pay more attention to the importance of forensic testing, and foreign assistance has been solicited, and provided, with increasing frequency. Joint investigations into the Bali bombings by the Australian Federal Police (AFP) and Indonesian National Police, for example, resulted in 33 convictions (Ramakrishna, 2004). Of particular note is the July 3, 2004, opening of the Australia-Indonesia Jakarta Centre for Law Enforcement Cooperation (JCLEC). A collaborative effort, JCLEC supplies Australian forensic experts to local investigations and provides education in forensic sciences to local officers. Nonetheless, local authorities must still rely upon foreign colleagues for some testing capabilities—for example, those to analyze chemical, biological, radiological, or nuclear signatures should such incidents occur.[28]

Countertechnology Responses by JI and Its Affiliates

JI appears to recognize the importance of denying police investigators forensic evidence that can be used to identify, track, and successfully prosecute members involved in bombings. In the 2002 Bali attacks, the group attempted to deface the chassis numbers of the explosives-laden minivan by changing its zeros to sixes, and it has since been using explosives in quantities and compositions sufficient to remove any investigative value from a detonation's remains (Nakashima, 2003; Sebastian, 2004; Turnbull, 2003). JI's shift to the use of suicide bomb-

---

[28] Personal interviews with local and Australian intelligence officials, the Philippines (March 2005) and Indonesia (April 2005).

ers has similar benefits, eliminating the chances of an attacker being captured alive and able to provide authorities with compromising intelligence (Yusuf, 2003).

Conclusion

Although information acquisition, preventive action, denial, and investigation technologies have improved Southeast Asia's ability to combat JI, their impact—individually and cumulatively—has been limited. Table 3.1 summarizes technologies that have been

Table 3.1
Jemaah Islamiyah Technological Innovations:
Purpose and Intended Mitigation of Government Countermeasures

Innovation	Purpose	Intended Mitigation of Government Countermeasures
Leveraging training facilities of other terrorist organizations	Allow training in lower risk areas	Overhead monitoring by surveillance assets
Using text-based messaging capabilities	Communication security	Government monitoring of voice communication
Limiting length of communications	Communication security	Government monitoring of voice communication
Using anonymous and disposable SIMs in mobile phones	Break identifying link between phone and operative	Government monitoring of voice communication
Using couriers to deliver messages	Communication security	Government monitoring of voice communication
Shifting among transport modes and border crossing	Avoid well-secured sites and crossings	Government border controls
Falsifying documents and using operatives who break profile	Deceive information-gathering efforts at borders	Government border controls
Using disguise and deception	Operational security	CCTV monitoring of targets and public places

Table 3.1—Continued

Innovation	Purpose	Intended Mitigation of Government Countermeasures
Disaggregating weapons and transporting in small pieces	Avoid detection in transit	Weapon detection technologies
Packing explosives with foodstuffs and masking with pungent odors	Avoid detection in transit	Weapon detection technologies
Triggering multiple explosions in quick succession	Act faster than government can deploy countermeasures to detonation signals	Cell phone jamming to prevent weapon detonation
Using redundant detonation mechanisms	Provide alternatives if some modes are jammed or circumvented	Cell phone jamming to prevent weapon detonation
Routing financial transactions through informal modes or permissive nations	Allow functioning of financial support structure	Government efforts to intercept funding to group
Using many small-scale transfers of funds rather than large single transfers	Circumvent controls put on large money flows	Government efforts to intercept funding to group
Shifting to soft targets	Avoid hardening measures at high-profile target sites	Protection at desirable targets
Increasing scale of bombs	Overwhelm hardening measures at high-profile target sites	Protection at desirable targets
Modifying operations in an attempt to penetrate target defenses	Avoid hardening measures at high-profile target sites	Protection at desirable targets
Destroying forensic evidence through preplanning or design of weapon systems	Maintain operational security	Government forensic science analytical capabilities

deployed against JI and the countertechnology strategies the group has implemented in response. Despite government efforts, JI has maintained a broad recruiting base, access to protected training facilities, and considerable financial support from multiple sources.

Southeast Asia, further, with its remote and inaccessible geographies, expansive coastlines, and densely populated cities, provides JI a highly permissive arena in which to operate. Persistent disparities in visa requirements and technical capabilities across states allow the group to circumvent the region's most threatening defenses and to exploit its most attractive vulnerabilities. Significant improvements in state systems are unlikely to occur in the near term, placing a premium on multilateral cooperation on activities ranging from the patrolling of borders, to the institutionalization of compatible banking regulations, to the sharing of intelligence.

Domestically, CCTV cameras and the hardening of targets are useful means through which to increase the resource burden that JI must accept in order to mount an operation. Neither, however, will necessarily prevent JI from pursuing targets it perceives to be of high value. Given the organization's demonstrated willingness and ability to adjust its strategies and tactics, it is more likely that JI will eventually counter such measures by increasing the surreptitiousness of its activities, by improving the design and yield of its explosives, or by innovating new means of delivering weapons to desired targets.

Nonetheless, Southeast Asia's experience with JI does suggest that current technologies can increase the direct risks and costs that a group must incur in order to carry out an attack, even if the technologies cannot prevent that attack from occurring. JI's operational history demonstrates that it has had more difficulty finding work-arounds for low-technology, on-the-ground deployments than it has for highly sophisticated systems. Hardened targets, CCTV, and weapon detection technologies have compromised operations in progress, curtailed casualty rates, and increased the likelihood of apprehending the attacker. Overhead surveillance, communication intercepts, and detonation-thwarting technologies, by contrast, have had little success in restraining the group from mounting and effecting attacks—counterterrorism professionals in the region emphasize that human intelligence remains the most important tool for prevention.[29]

---

[29] Personal interviews with local officials, Indonesia, the Philippines, Singapore, and Thailand (March–April 2005).

CHAPTER FOUR

Liberation Tigers of Tamil Eelam

Introduction

Sri Lanka is a small island located off the southern coast of India. The country is roughly 65,610 square kilometers and is split into nine administrative districts.[1] It has a total population of slightly more than 18 million, three quarters of which is Sinhalese. Roughly 17 percent of the population is Tamil, with Moors, Burghers, Eurasians, and Malays constituting the bulk of the rest. Approximately two-thirds of the population is Sinhalese-Buddhist; Hindus and Muslims together account for 14 percent of the total, with the residual made up of Roman Catholics and other Christian groups (Ministry of Finance and Planning, 1998).

The Sinhalese are mostly concentrated in the southern, western, and central parts of Sri Lanka, having originally come to the island from India. The roots of their civilization are, thus, largely Indian, although they have been influenced by other cultures, including the Portuguese, English, and, to a lesser extent, the Dutch, Burmese, and Thai. The bulk of the Tamil population is located in the drier northern and eastern provinces of Sri Lanka and is split into two distinct groups: the Jaffna Tamils, who are mainly descendants of tribes that first arrived on the island well over 1,500 years ago, and the Indian Tamils, who originate from plantation workers, brought to the island

---

[1] These include the following provinces: Western, Southern, Uva, Eastern, Central, Sabaragamuwa, North Western, North Central, and Northern.

by British tea planters during the 19th and early 20th centuries (de Silva, 1996).

The principal internal conflict that has plagued Sri Lanka since the country gained its independence from the United Kingdom in 1948 revolves around the Tamil campaign for a separate Eelam state in the north and east of the country. The roots of this struggle date back to Tamil resentment of government "standardization" policies—particularly those relating to quotas for admission to universities,[2] introduced during the 1950s, 1960s, and 1970s in an attempt to rectify ethnic discrimination to which the majority Sinhalese community claimed they had been subjected under British colonial rule.[3] Reacting to a series of discriminatory moves that were designed to ensure Sinhalese domination of the country's main educational institutions and bureaucracy, several opposition Tamil groups banded together in 1972 to form the Tamil United Front (TUF). At first, the group campaigned simply for linguistic, ethnic, and religious equality throughout the country. However, by the mid-1970s, the TUF had become associated with a far more hard-line, nationalist stance, largely in reaction to the continued refusal by the Colombo government to grant even limited concessions to the Tamil minority. In 1976, the TUF renamed itself the Tamil United Liberation Front (TULF), contesting the 1977 Sri Lankan elections on a mandate that called for the creation of a fully independent Tamil state of Eelam (see Gunaratna, 1998; Joshi, 1996; Thackrah, 1987; *The Europa Yearbook*, 1998; and Thomas, 1994).

Although the TUF, and later the TULF, were prepared to agitate for independence through the accepted political channels of the Sri Lankan state, a hard-core element of the organization viewed extra-constitutional violence as the only means by which Tamil nationalist

---

[2] Education is highly prized among the Tamils; the introduction of university admission quotas has been identified as one of the principal factors that originally drove the community's youth to militancy (email correspondence between author and Sri Lankan intelligence official, May 2005).

[3] The Sinhalese claimed that, under colonial rule, the British had practiced an explicit pro-Tamil policy in an attempt to marginalize Sinhala independence designs and aspirations. This, it was argued, had placed the Tamils, who constituted only 12 percent of the population, in a position of disproportionate power and authority at the time of independence.

objectives could be achieved. During the 1970s, these militants formed a variety of underground guerrilla organizations dedicated to armed struggle against the Colombo government, using the TUF and TULF more as a secondary vehicle for political representation (in much the same way as PIRA does with Sinn Fein). Initially, 35 militant groups were created, although five quickly achieved dominance—one of which, LTTE, or the Tigers, has remained at the forefront of the Tamil civil war in Sri Lanka.[4]

Led by Velupillai Prabhakaran, LTTE has waged a bitter struggle for Tamil self-determination in Sri Lanka's northern and eastern provinces during the past four and a half decades.[5] During this period, the Tigers have gained a reputation as one of the most sophisticated and deadly terrorist insurgencies in the world, successfully driving Colombo to the negotiating table in February 2002 and effectively forcing the government to accept terms for a cease-fire that have since allowed the group to set up a mini Eelam state covering roughly 15 percent of the country's geographic territory.[6]

---

[4]   The other four groups included the Tamil Eelam Liberation Organisation (TELO), the People's Liberation Organisation of Tamil Eelam (PLOTE), the Eelam People's Revolutionary Liberation Front (EPRLF) and the Eelam Revolutionary Organisation of Students. LTTE itself was originally named the Tamil New Tigers.

[5]   The ideological basis of the LTTE separatist struggle is set out by the so-called Thempu Principles, which affirm recognition of

- the Tamils as a nation
- the existence of an identified homeland for the Tamil people
- the right of the Tamil people to self-determination
- the right of the Tamil people to a separate citizenship
- the fundamental right of all Tamils to look upon the north and eastern provinces of Sri Lanka as their own country.

[6]   Personal interview with Western diplomat, Sri Lanka (December 2002). The cease-fire was brokered by Norway on February 22, 2002, and has since led to several rounds of talks between Colombo and LTTE. At the time of this writing, the Tigers had put forward their own blueprint for home rule and have given explicit warning that, if this proposal is not taken seriously, they will again take up arms against the government. Many in Colombo believe that the Tiger leader has no interest in peace and is merely using the current period of relative stability to rearm, recruit additional cadres, and consolidate control over the north. (personal

To be sure, adroit employment of guerrilla tactics and Colombo's own military incompetence[7] have been important factors in accounting for the Tigers' success. However, a critical element that has helped to amplify these battlefield modalities has been the efficient and consistent use of unconventional martyr attacks that have ranged from selective assassinations to large-scale assaults undertaken against economically, politically, or militarily strategic targets. Involving thoroughly trained operatives who have proven their ability to act decisively on land and sea and incorporating unique and innovative intelligence and counterintelligence methods, it is these martyr operations that have, arguably, become one of the most infamous hallmarks of the Tamil ethnonationalist war as waged by LTTE.

This chapter assesses LTTE tactical development in the field of suicide terrorism, paying specific attention to modalities that have been adopted to circumvent government-instituted countermeasures against this particular form of nonstate violence.[8] It first provides an overview of the organization's land-based and sea-based martyr capabilities and

---

interviews with Sri Lankan intelligence and military officials, Bangkok and Colombo (May 2004 and April 2005). See also International Institute for Strategic Studies, 2004; "Peace Process Bogged Down in More Questions," 2004; and "Peace Talks," 2004).

[7]   In many ways, the Sri Lankan Armed Forces (SLAF) has yet to emerge as a professional force that truly understands the nature and type of war it has been fighting in Sri Lanka. The majority of commanders have never seen any action, with many promoted purely on the basis of time served or as a result of political connections, personal loyalties, and friendships. Compounding the situation is the wholly inadequate training and support that is given to regular soldiers. Indeed, some recruits have been dispatched to the front line after only four weeks of basic combat training, and troops regularly cite shortages in basic equipment such as modern assault rifles, ammunition, and field radio sets. The SLAF has also tended to rely on outdated doctrines that place a premium on taking and holding static lines of defense through maximum force as opposed to more nuanced (and relevant) counterinsurgency operations (personal interviews with Western diplomat and Sri Lankan military officials, Sri Lanka, May 2004).

[8]   Because of LTTE's size and the nature of much of its operations, a large fraction of its activities is more insurgent or military-on-military than terrorist in nature. Although LTTE has engaged in efforts to circumvent defensive measures in those activities, they are more relevant to a military, rather than a homeland security, context. As a result, the following discussion addresses the group's suicide terrorism operations, which restricts the scope of the examination to activities that are directly relevant to the homeland security focus of the study.

then discusses some of the more notable innovations the Tigers have enacted to blunt efforts by the Sri Lankan security forces to detect and otherwise prevent suicide operations from achieving fruition.

LTTE Suicide Terrorist Capabilities and Infrastructure: The Black Tigers and Sea Tigers

LTTE suicide strike teams are vested in two operational wings: the Black Tigers (BTs) and Sea Tigers (STs).

The Black Tigers. The BTs constitute LTTE's main suicide wing. The division currently has about 350 members who are invested with the responsibility for carrying out three main types of operation:

- those used primarily on the battlefield and directed against combat troops of the SLAF (Sri Lankan Armed Forces, not to be confused with the Sri Lankan Air Force) as well as military personnel and assets in rear areas or defined war zones
- those aimed at critical national infrastructure, civilians, and urban complexes such as railway stations, religious shrines, and banks
- those that target what LTTE expansively defines as VIPs, including elected leaders; prominent political figures; other high-level government functionaries; senior military, police, and intelligence commanders; and, occasionally, lower-ranking members of the security community whose competence has attracted the attention of the Tigers.[9]

The BTs receive perhaps the most demanding training of any LTTE unit, involving endless physical endurance, and psychological and indoctrination sessions. Prospective suicide bombers are taught where to place themselves and their bombs to be most effective and how to avoid detection. Those who successfully pass the "death course" are

---

[9] Personal interview with Sri Lankan intelligence official, Sri Lanka (May 2004). Because many of the second and third categories of operation take place in Colombo—and, at times, have caused significant numbers of collateral casualties among civilians—LTTE suicide attacks on nonmilitary targets are never claimed. Such a stance is designed to limit the negative political fallout that is often an inevitable by-product of the resort to martyr-based modalities.

subjected to the tightest security, their identity generally known only to the highest echelons of LTTE leadership.[10] The reason for this intensive regime is that, unlike groups such as Hamas, PIJ, and Chechen terrorists, LTTE reserves suicide bombers for the most important Tiger operations and missions (see appendix).[11]

All BTs are selected from "conventional" LTTE military ranks. Inductees are observed for an extended period of time—sometimes as long as five years—and chosen on the basis of their ability to blend into unfamiliar environments, their capacity to operate independently and think on their feet, and their perceived hatred of the enemy. Recruitment has, thus, tended to focus on appropriately skilled individuals who have also directly experienced some form of abuse or worse at the hands of the authorities.[12] BT intakes are small, generally numbering no more than 30 cadres at a time. Once nominated, trainees are given a new identity and separated from the wider LTTE community, which both avails the aforementioned requirement for operational security and gives instructors greater latitude to imbue further the ethos of martyrdom and self-sacrifice.[13]

BT training lasts for approximately eight to nine months (although it can extend to a year for cadres charged with assassinating especially high-profile targets) and is split into two components:

- Phase I focuses on honing physical and mental fitness and developing proficiency in a baseline set of skills such as bomb construction, weapon handling, vehicle and motorcycle driving, counter-surveillance, and anti-interrogation.

---

[10] During training, all Black Tigers are hooded, as are their instructors. Even when dispatched on joint missions, members will typically only learn the identity of others in their team immediately prior to the attack (personal interview with Sri Lanka intelligence official, Thailand, April 2005).

[11] Personal interview with Sri Lankan intelligence official, Sri Lanka (May 2004).

[12] Personal interviews with with Sri Lankan intelligence and military officials, Sri Lanka (May 2004).

[13] Personal interview with LTTE member, Sri Lanka (February 2003).

- Phase II emphasizes specialized, mission-oriented training. During this part of the course, BT instructors separate cadres according to designated operations—some to prepare for selective assassinations, others to carry out strategic assaults and attacks against the SLAF.

Those assigned to kill VIPs receive technical or career instruction as well as reading and writing instruction, while those who are retained for more complex military and critical infrastructure strikes are taught advanced covert penetration techniques and decoy methods (the latter typically designed to maximize the number of casualties in a defined "kill zone"). In both instances, operatives continually practice simulated missions, either by using models of designated targets or performing test runs to evaluate the effectiveness of extant security measures and procedures.[14]

In the context of civilian violence, BT attacks have largely focused on VIP assassinations and vehicular explosives. In the former case, the "hit team" usually involves the dispatch of a single operative who is supported by a handler and a Tigers' Organisation for Security Intelligence Service (TOSIS) unit. The typical modus operandi for the martyr is to detonate a suicide vest once he or she is within the immediate vicinity of the target. In most cases, those selected for assassination are high-profile individuals who have been identified as posing a direct strategic threat to the group, meaning that the Tigers typically take great stock in ensuring a successful kill.[15]

Truck and van explosive delivery devices have also been frequently used, mostly to destroy buildings that have either symbolic or strategic importance. Some of LTTE's most audacious suicide bombings have employed this method, including, notably, strikes against a Tamil University taken over by the SLAF in 1987, the Joint Operations Center at the Ministry of Defense in 1991, the Sri Lankan Central Bank in 1996, and the Colombo World Trade Center in 1997 (see appendix).

---

[14] Personal interviews with former BT member, Sri Lanka (May 1998) and with Sri Lankan intelligence official, Thailand (April 2005).

[15] Personal interviews with military officials, Sri Lanka (May 2004).

The incident in 1996, which left 91 people dead and more than 1,400 injured, remains the most destructive act of terrorism to have ever been carried out in Sri Lanka and is generally recognized as a textbook example of long-range strategic planning, logistical support, and operational security.[16]

LTTE employment of suicide terrorism is unparalleled in terms of its effectiveness. The Tigers are not only unique in being the only substate group to have killed two heads of state (Prime Minister Rajiv Gandhi in 1991 and President Ranasinghe Premadasa in 1993; see appendix), their consistent and deadly use of martyrdom is widely believed to have been one of the main factors that drove Colombo to the negotiating table in 2002.[17] As one former Sri Lankan foreign service officer and ambassador remarked, "This is an example where terrorism has succeeded. We have been cowed. We have been intimidated by suicide terrorism. It is that simple. The fear caused by this tactic has made us cave into them."[18]

The Sea Tigers. The STs form LTTE's maritime wing. The unit's current strength is estimated to be between 3,000 and 4,000—some 2,100 Sea Tigers are thought to have perished as a result of the massive tsunami that struck Sri Lanka in December 2004[19]—who are organized into operational divisions covering engineering, maintenance, and communication personnel; underwater demolition teams; naval trainers; and suicide strike forces. Tiger marine facilities and bases are

---

[16] Personal interview with Sri Lankan intelligence official, Thailand (April 2005). See also Yapa (1996) and Jayasinghe (1996). The attack involved the predeployment of a BT suicide team some 90 days before the operation. Members were thoroughly versed in the nature of their mission, undertaking countless hours of surveillance to assess the overall vulnerability of the venue and the best means for overcoming extant security measures. Moreover, to minimize the possibility of the mission being compromised, all information was compartmentalized and transferred on a need-to-know basis only.

[17] Personal interview with Sri Lankan intelligence official, Thailand (April 2004).

[18] Personal interview with former Sri Lankan foreign service officer, Sri Lanka (February 2003).

[19] The STs were severely impacted by the tsunami, largely because most of the unit's vessels were berthed at the time the tidal wave struck. In the words of one intelligence source, "In seven minutes to 10 minutes it was all over—radar and communications facilities, munitions dumps and dry docks were all basically devastated" (Davis, 2005a, p. 39).

strung along the northeastern coast from Chundikulam in the north to areas near to and south of the government-held port of Trincomalee (Davis, 2005a, p. 39).

Unlike their land-based counterparts, ST martyrs are not specially trained, tending to consist of wounded militants who volunteer to undertake suicide missions as a last "hurrah" to the group (the one exception being underwater combat divers, who, as indicated below, rarely are called on to undertake "self-sacrifice" operations). All STs have extensive knowledge of the maritime environment, are highly experienced in sea-based operations, and are fully adept at covert, surprise attacks against surface vessels. Moreover, the willingness of injured STs to die rather than take up land-based, logistical duties is testament to the unbending loyalty of these fighters—both to Prabhakaran personally and to the Tamil cause in general.

ST suicide strikes typically involve the use of explosive-laden boats that are rammed into surface frigates that have been singled out and surrounded by hunter "wolf packs." Attack craft are usually crewed by one cadre (although for more important missions two mariners may be used) and are shallowly hollowed out in the fashion of a shoe and constructed from lightweight fiberglass material, both to maximize their speed and maneuverability and to reduce their radar cross-section.[20] Vessels are typically rigged with between 10 and 14 Claymore mines that are connected in a circuit to three booster charges, weighing up to 21 kilograms each. Boats also often have special penetration steel spikes that are attached to their bows, which are designed to puncture hardened hulls of targeted vessels on impact. This style of attack has been highly effective in amplifying the destructive force of resulting shock waves, ensuring that even large-scale combat ships will sink following the detonation of explosive packs.[21]

ST martyr operations have been as decisive as those carried out by the BTs. Since 1990, LTTE has carried out more than 40 suicide attacks at sea, the basic aim of which has been to disrupt the mobil-

---

[20] See, for instance, "LTTE Suicide Kit Assembly Plant in Dehiwala Raided" (2001).

[21] Email correspondence with Sri Lankan intelligence official (May 2005); Chalk and Hoffman (2005).

ity of Sri Lanka Navy (SLN) patrols off the northeast coast (a critical smuggling conduit for LTTE arms procured from overseas).[22] According to Sri Lankan sources, most of the attacks have been effective in significantly damaging, if not sinking, naval surface ships. According to one senior retired SLN officer, fear of being caught in one of these strikes has been one of the main factors accounting for reduced recruitment into the SLN.[23] It is also salient to note that LTTE conducted assaults similar to that undertaken by al Qaeda against the USS *Cole* (a U.S. Navy destroyer that was hit while anchored at the Port of Aden in October 2000) as far back as 1995.[24] This suggests not only that the STs are some years ahead of al Qaeda in terms of seaborne capabilities but, more importantly, may be serving as a critical benchmark guiding developments in the wider area of maritime terrorism.

LTTE Innovation in Suicide Terrorism Technology: Responses to Government-Instituted Countermeasures

LTTE's innovation in suicide technology has been driven by a combination of independent initiative, reflecting a highly active internal research and development program that has evolved under the auspices of the group's chief explosives expert, Wedi Dinesh, as well as responding to government countermeasures. Indeed, it is Dinesh who first thought of the idea of using clothing to secret bombs, a technique that is now standard practice for groups across the Middle East and Asia. Initially the emphasis was on specially designed denim shorts that were capable of carrying a payload of between 1.5 and 1.6 kilos of explosive material. When these proved too small for larger-scale attacks, Dinesh refined the delivery mode, manufacturing suicide vests containing a pouch of steel ball bearings that was placed between two explosive slabs of 2.5 kilos each. It was also under his tutelage that the Tigers developed various automated and nonautomated means for carrying

---

[22] Personal interview with Sri Lankan military intelligence, Sri Lanka (May 2004).

[23] Personal interview with former SLN officer, Sri Lanka (May 1999).

[24] Personal interview with Western diplomatic official, Sri Lanka (May 2004). For example, LTTE attacked the SLN gunboats *Suraya* and *Ranasuru* in 1995, destroying both ships (see appendix).

out martyr strikes, ranging from cars and trucks to motorbikes, boats, "tuk tuks" (small motorized taxi vehicles), and bicycles.[25]

As noted in the introduction, technologies that are designed to disrupt terrorist activities fall into the following five categories: (1) those that acquire or manage information; (2) those supporting preventive action; (3) those that are aimed at denying the terrorists' ability to attack targets; (4) those that are aimed at responding to the effects of an attack; and (5) those that are aimed at investigating after the fact. In the context of BT and ST attacks, the main thrust of Sri Lankan mitigation efforts fall into categories 1 and 3.

Information Acquisition and Management

Gathering information on LTTE activities is a significant element of Sri Lankan efforts to defeat suicide bombing operations and has been the focus of significant competition between the group and government organizations.

Technologies Deployed

In terms of information collection, Sri Lankan security forces place a premium on insider intelligence procured directly from the Tigers' ranks. Considerable emphasis is also placed on intercepting communications to and from the group's intelligence wing, TOSIS, which, apart from the senior leadership, is perhaps the only subcomponent of the Tigers' institutional structure that has a detailed overview of the identity of individual members of the BTs and STs and their proposed attack venues. This data is collated, assessed, and cross-referenced by security forces with existing information both to build profiles of "typical" Tiger suicide operatives (which facilitates the general process of

---

[25] Personal interview with Sri Lankan terrorism expert, Singapore (April 2005). The importance that Prabhakaran accords to Dinesh was made apparent in the late 1990s, when the explosives technician was wounded in a Sri Lankan air force raid. According to the interviewee, the first thing the LTTE leader asked on hearing the news was: "Is his brain OK?"

police profiling—described below) and attendant strike patterns and, ideally, to preempt planned attacks.[26]

To restrict the movement of suicide operatives and preempt single-cadre assaults and assassinations (a signature trait of the BTs), the government has instituted a multidimensional approach that is closely modeled on the Israeli concept of layered security.[27] This particular technique takes the form of a defensive cone consisting of three concentric circles that seeks to place successive barriers in the path of the martyr: the nontarget area, the pretarget area, and the target area. The aim is to identify bombers and attendant handlers or scouts in the outer rings and then to progressively funnel them inward, where they can be isolated and engaged in a place and time of advantage to the authorities. At all points in the matrix, police are trained to be in a "hunter" rather than a "fisherman" mind-set—actively seeking out, tracking, and observing their "prey" rather than waiting passively until something demonstrably threatening occurs.[28] Integral to the Sri Lankan version of layered security is the deployment of agents specially versed in reading body language and profiling potential suicide bombers and handlers or scouts. Most of these officials are stationed in the outer rings of the defensive cone and charged with passing on information to advance security teams mandated with sweeping and securing approaches to the main target area.[29]

Sri Lankan security forces have also instituted a range of initiatives applying detection modalities that are designed to identify bombs and other improvised explosive devices (IEDs). Through experience, the security forces have learned not to rely on a purely technological approach, but to incorporate a combination of search methods that typically embrace body "pat downs," sniffer dogs, metal and vapor detectors, and x-ray scans. Taken together, this multidimensional format is

---

[26] Personal interviews with intelligence and military officials, Sri Lanka (May 2004) and Thailand (April 2005).

[27] Personal interview with Sri Lankan intelligence official, Thailand (April 2005).

[28] Personal interview with Bruce Hoffman, RAND Corporation, Arlington, Va. (June 2005).

[29] Personal interview with Sri Lankan intelligence official, Thailand (April 2005).

estimated to reduce extant risk levels to as low as 8 percent—compared to 85 percent when a single technological technique is used (see Table 4.1).[30]

Countertechnology Responses by LTTE

As noted above, the defensive cone has been used primarily to preempt LTTE single-cadre suicide strikes, which, in most cases, involve assassination attempts against VIPs and government and military officials who have been identified as posing a direct strategic threat to the group and its objectives. Because these individuals are deemed especially high-priority targets, the Tigers go to considerable lengths to ensure that "hits," once initiated, are successfully carried out.[31]

To this end, the group has developed several procedures to minimize the risk of a suicide mission being interdicted by the Sri Lankan security forces before completion. As noted above, all BTs are thoroughly trained and never dispatched to a target area that has not been subjected to in-depth reconnaissance, which may extend for up to three months.[32] To minimize the risk of early interception, martyrs employ a range of countersurveillance techniques (such as never adhering to one daily pattern and ensuring that their behavior is "consistent" with

Table 4.1
Estimates of the Efficiency and Effectiveness Matrix of Search Modalities According to Sri Lankan Intelligence Sources

Type of Search	Relative Efficiency
Detectors	Effective 15% of the time
Dogs	Effective 68% of the time
Human hand	Effective 72% of the time
Detectors + dogs + human hand	Effective 92% of the time

[30] Personal interview with Sri Lankan intelligence official, Thailand (April 2005).

[31] Personal interviews with military officials, Sri Lanka (May 2004).

[32] Surveillance is used to determine two main things: (1) How vulnerable is the target? and (2) What sort of attack would be most appropriate given the extant security procedures that are in place?

their specific operational context) and typically will not don their suicide vests until they are within 1 kilometer of the strike location (the device itself having been either predeployed or transported to the target venue by a third party). Finally, to overcome security force profiling, the BTs have emphasized the recruitment of operatives who do not readily conform to stereotypical images of suicide bombers, including women and, allegedly, children.[33]

In addition, LTTE has invested a significant amount of time and resources in fine-tuning BT explosive packs to ensure that, having passed through the defensive cone, operatives will not expose their position by attempting to detonate a faulty device. All suicide vests are fully tested[34] and checked before deployment, and each is outfitted with an LED indicator to verify that the circuit and power supply is intact and operational. Body suits are also designed with built-in redundancy and "fail-safe" systems involving secondary and, occasionally, tertiary trigger switches.[35] According to one highly informed source in Colombo, the Tigers now routinely use suicide vests that can be automatically detonated simply when the wearer raises his or her arms. These have been used with deadly effect in assassinations involving physical embraces of the selected victim and the placement of wreaths or garlands over the victim's head.[36]

The BTs have employed a range of methods to defeat Sri Lankan detection modalities. For assassinations, the BTs have increasingly turned to female operatives, who, at least initially, were less likely to be viewed as potential suicide bombers.[37] Moreover, given the extreme modesty that transfuses Sri Lanka's Buddhist society, women are gen-

---

[33] Personal interviews with Sri Lankan intelligence official and LTTE expert, Thailand and Singapore (April 2005).

[34] Suicide vests are usually tested on animals until they reach the desired results in terms of explosive power and the direction of shock waves (personal interview with Sri Lankan terrorism expert, Singapore, April 2005).

[35] Personal interview with Sri Lankan intelligence official, Thailand (April 2005).

[36] Personal interview with Western diplomatic official, Sri Lanka (May 2004).

[37] Overall, roughly a third of all suicide missions in Sri Lanka have been conducted by female BT cadres.

erally not subjected to the same type of comprehensive body searches that are typically used for men, which means they have a higher likelihood of bypassing random checks. BT "body bombs" are also airtight and constructed with wax-coated wiring to reduce the risk of telltale vapors being picked up by sniffer dogs.[38] The latest technological innovation has been the development of a heart-shaped suicide vest, which is worn by a woman and which is designed to hold explosive slabs in two bra cups surrounding the breasts. Charges are detonated via one of two triggering mechanisms: one that runs up the center of the torso and one that is placed under the right or left armpit. Sources in Southeast Asia stress that these suits have been deliberately manufactured to bypass body hand searches (which, given the modesty noted above, do not routinely emphasize the bosom) and are being readied for an intensive campaign of suicide assassinations in the event that the current peace process in Sri Lanka fails.[39]

Innovation to defeat detection technologies has been just as apparent with regard to vehicular bombs. Initially, explosives were placed in the side panels of cars and trucks. These devices proved to be susceptible to routine searches and were often discovered simply as a result of knocking on doors or structural wings and listening for whether the resulting timbre was "solid" sounding (which would strongly indicate that the paneling had been removed and the space packed with foreign material).[40]

In reaction, LTTE designed bombs that could be hidden in specially modified fuel tanks that are linked to a booster charge (generally TNT) and a detonation cord attached to a trigger switch located in the driver's cab. To minimize the risk of explosives being discovered via the insertion of "dipper" probes, gasoline reservoirs are lengthened and then retrofitted with an artificial separation wall, forming a separate base compartment in which the IED would be placed. Like suicide vests, all bombs are airtight and wired with wax-coated circuitry

---

[38] Personal interview with Sri Lankan intelligence official, Thailand (April 2005).

[39] Personal interviews with Sri Lankan intelligence and security officials, Thailand and Singapore (April 2005). See also "Expert Warns of Breast Type Suicide Kits" (2005).

[40] Personal interview with Sri Lankan intelligence official, Thailand (April 2005).

to prevent the emission of betraying vapors; as an added precaution against detection technologies, payloads are also wrapped in plastic.[41]

Over the years, LTTE has further refined the technical procedures of attacks in which explosives are loaded on trucks. According to Sri Lankan intelligence officials, most IEDs now take the form of C4-TNT Composition B explosive bundles that are surrounded by a nonmetallic outer layer. The devices are linked with a single detonation cord and packed into a false bottom that runs the length of a vehicle above its chassis. These storage compartments are lined with mint and other canine "detracting" spices such as cardamom, pepper, and cinnamon. The combined effect has been the development of a bomb that is now essentially immune to casual visual inspections and detection by sniffer dogs and automated scanners.[42]

Besides defeating operational and technical modes aimed at detecting its suicide operatives, LTTE also has demonstrated a degree of innovation and sophistication in subverting government intercepts of its communications. Thanks to training provided by India's Research and Analysis Wing (the agency charged with advancing Delhi's clandestine foreign policy goals) during the 1980s,[43] the group has a well-founded grasp of disinformation techniques that have been used to mask attack plans and strategies.[44] To avail secure lines of contact between operational cadres and wider support and intelligence teams, the group also only uses prepaid cell phones, which cannot be traced. Most SIM cards

---

[41] Personal interview with Sri Lankan intelligence official, Thailand (April 2005).

[42] Personal interview with Sri Lankan intelligence official, Thailand (April 2005).

[43] India played an important role in militarily backing LTTE (as well as other principal militant organizations such as PLOTE, TELO and EPRLF) during its formative years. By the mid-1980s, it is estimated that some 20,000 militants had received insurgent training in India, most of which was conducted in dedicated paramilitary camps located in Tamil Nadu, Delhi, Bombay, Vishakhapatnam, and Chakrata (the country's premier military academy). Support for LTTE (and other Tamil organizations) was curtailed in the second half of the 1980s, however, on account of growing law and order problems that had been created by the presence of armed Tamil militants in the south, which were compounding the already serious socioeconomic strains that had been brought about by the number of refugees fleeing across the Palk Strait from Jaffna. For further details of this period, see Gunaratna (1997), Abraham (1998), Dixit (1998), and Tilakaratna (1998).

[44] Personal interviews with intelligence and military officials, Sri Lanka (May 2004).

are bought from private vendors in small towns and outlying suburban districts, largely because they do not typically request and keep accurate records of a purchaser's identity and place of residence at the point of sale. Although not as convenient as larger, nationwide wireless phone companies, calls made from phones using these cards are essentially invisible in terms of ownership and, therefore, constitute an ideal medium for coordinating and executing martyr (and other terrorist) assaults.[45]

LTTE is also known to use Thuraya satellite phones. These devices operate via an exclusive signal that is transmitted through the United Arab Emirates, which, again, makes them extremely difficult to monitor. Sri Lankan intelligence officials concede that it is unlikely that Colombo will have the technology to eavesdrop on Thuraya phones for several years, leaving LTTE with an internal communication mode that, at least for the short to medium term, will remain insulated from government interception.[46]

Finally, TOSIS has fine-tuned the art of discursive code-writing to defeat government counterintelligence operations. Operational memos and orders to suicide cadres are frequently contained in the text of ordinary magazines, the content of which is compiled from separate words drawn from pages known only by the intended recipients. An additional technique involves the use of letters that are impossible to read without a specially cut Slidex™-type chart (which hides all text other than that pertinent to the message).[47] This technique has been used to convey a broad array of information relevant to an intended target site, including extant security procedures, alternative penetration routes, vulnerable blind spots, and, in the case of assassinations, the location of predeployed IEDs.

---

[45] Personal interviews with intelligence officials, Thailand and Singapore (April 2005). There has also been some speculation that LTTE has sought to procure SIM cards from Thailand and the Philippines—which work across existing networks in Sri Lanka, again because there is no formalized procedure in place for recording the identity of purchasers.

[46] Personal interview with Sri Lankan intelligence official, Thailand (April 2005).

[47] Personal interview with Sri Lankan intelligence official, Sri Lanka (May 2004).

Denial

Target hardening has been a current theme of Sri Lankan efforts to mitigate BT and ST attacks. Technologies deployed have included both procedural approaches as well as more traditional strategies for protecting and hardening targets.

Technologies Deployed

Close protection teams generally accompany high-profile government officials, politicians, and military or intelligence officials whenever they travel or appear in public. For individuals deemed to be especially at risk, a personalized VIP staff or aid team will also be dispatched to ensure that there is absolutely no deviation from predetermined security plans.[48]

To guard against vehicular attacks against prominent financial, commercial, and government buildings and transportation hubs, various target-hardening procedures have been adopted. Similar to those safeguards employed in other suicide terrorist–prone countries, these typically include some or all of the following:

- installation of outer perimeter defenses such as vehicular monitoring stations, speed bumps, zig-zag barriers, and surface girders constructed from rail tracks[49]
- closure of car parks located within a preconfigured blast radius (typically only used for high-profile buildings such as the president's and prime minister's official residences, foreign embassies, and Parliament)
- institution of various internal security procedures covering identity verification; logged entry, exit, and meeting details; and sanitized forward holding areas for visitors.[50]

---

[48] Personal interview with Sri Lankan intelligence official, Thailand (April 2005).

[49] Written correspondence between author and Sri Lankan security expert (May 2000).

[50] Personal interviews with Sri Lankan and Western officials, Sri Lanka and Thailand (May 2004 and April 2005).

For seaborne assaults, the SLN has moved to reinforce the super-structure of surface frigates patrolling waters close to Tiger littoral bases, as well as to equip these vessels with radar technology to detect incoming attack craft. Stringent security procedures have also been put in place at the port of Colombo—the country's main commercial outlet—incorporating random physical checks of all dock personnel, constant monitoring of surface waters by the marine police, and, according to one Western diplomatic official in the region, the laying of mines to protect the mouth of the harbor from unauthorized ST incursions.[51]

Countertechnology Responses by LTTE

LTTE has instituted several means to overcome target hardening. In the realm of assassinations, most effort has been directed at infiltrating suicide cadres directly into the ranks of close protection and VIP staff and aid teams. A major component of BT training involves lessons on how to act, talk, and think in a wide range of environments—skills that have been used to infiltrate a broad swath of venues, including government bureaucracies. This regimen's effectiveness was perhaps best demonstrated in the 1993 murder of Ranasinghe Premadasa, who, as noted in the appendix, was killed by a deep penetration mole who had been on the presidential staff for several years.

Innovation has also been apparent with regard to seaborne attacks. To defeat SLN radar scans, for instance, ST suicide teams typically sail in close formation, closely hugging the coastline. The technique is designed to mask the signature of individual attack craft, both by avoiding radar signals altogether or, failing this, to give the impression of one large vessel. The tactic is based on the same procedure that combat air wings use to avoid aerial surveillance and, according to Sri Lankan intelligence officials, has been highly effective in facilitating covert approaches and surprise strikes against naval frigates, destroyers, and transporters.[52]

---

[51] Personal interview with Western official, Sri Lanka (May 2004).

[52] Personal interview with Sri Lankan intelligence official, Thailand (April 2005).

In addition to surface vessels, LTTE has emphasized refining underwater strike modes to circumvent harbor naval patrols and point-of-entry explosive barriers. The group has a dedicated combat diving cadre at its disposal—reputedly trained by ex-members of the Norwegian military—who are deployed for both conventional and nonconventional missions. Because of the investment made in those individuals to develop their unique skills, martyr operations are definitely more the exception than the rule for the ST diving cadre. However, the group has occasionally been prepared to use suicide divers to undertake stealth attacks against docked warships and other high-value maritime assets, usually by requiring a frogman to self-detonate submersible charges that are attached to a ship's hull or suspended from its propeller shaft.[53] Moreover, the Tigers are thought to have developed minisubmarines for covertly transporting martyrs inside strategically and commercially important harbors such as Colombo and Trincomalee.[54] Revelations that LTTE was moving in this direction first broke in 2000, when a partially completed minisubmarine prototype was discovered at a Tamil-owned shipyard in Phuket. According to informed sources, the five-meter vessel, while rudimentary, was capable of remaining submerged for up to six hours (at speeds of about five knots) and could very well have served as the blueprint for the more advanced versions that the STs are now alleged to possess.[55]

---

[53] Personal interview with Sri Lankan terrorism expert, Singapore (April 2005). In the words of this expert, Prabhakaran always has a "heavy heart" when authorizing underwater suicide missions on account of the inevitable "skill-loss" these operations entail.

[54] Sri Lankan sources also believe that the move to develop submarines was driven by the Navy's purchase of new-generation Dvora fast-attack craft at the end of the 1990s, which were proving effective against the Sea Tigers' surface ships.

[55] Personal interviews with Sri Lankan intelligence officials, Sri Lanka (May 2004). See also "Lanka Suspects Submarine in Thailand to be LTTE's" (2000) and Davis (2000). It is not currently known whether LTTE has been able to successfully introduce submarines into its overall battle armory.

Conclusion

LTTE has devoted substantial resources and time to defeating government-instituted tactics aimed at disrupting BT and ST suicide strikes. Table 4.2 summarizes the various LTTE suicide technological and communication innovations discussed above, the purposes they are intended to serve, and, where relevant, the government-instituted countermeasures they are designed to defeat. This intensive action-reaction dynamic across a broad range of technologies bears stark testimony to the importance the Tigers accord martyrdom, the use of which (unlike in organizations such as Hamas, PIJ, and al Qaeda) is reserved only for the group's most important missions. Indeed, virtually every initiative put forward by Colombo has been met with a response that is equal, if not superior, to the countermeasure being enacted. The comprehensiveness and effectiveness of these efforts can be gauged by the fact that some 80 percent of BT and ST operations are believed to have been instrumental in achieving their primary aims—a success rate unparalleled by any other group currently in existence.[56]

LTTE's conviction on the utility of suicide terrorism—and the need to ensure its continued integrity—owes much to the influence of the group's leader, Prabhakaran. The self-styled Tiger supremo has consistently held that martyrdom is a decisive force multiplier that is critical to the attainment of an independent Tamil Eelam.[57] Guaranteeing the success of BT and ST operations has thus emerged as a priority of the highest order, which, at least in the eyes of Prabhakaran, simply cannot be compromised by the state's countervailing activities.

Interestingly, however, there is one area to which the LTTE leader has chosen not to devote concerted attention: overcoming government target hardening of strategically or symbolically significant buildings. Although it is impossible to know exactly why this is the case, it may well be because defeating protective barriers such as vehicular

---

[56] Personal interview with Western diplomatic official, Sri Lanka (May 2004).

[57] In Prabhakaran's words, "With perseverance and sacrifice, Tamil Eelam [may] be achieved in 100 years. But if we conduct Black Tiger [suicide] operations, we can shorten the suffering of the people and achieve [this objective] in a shorter period of time" (Gunaratna, 2000, pp. 5–6).

Table 4.2
Liberation Tigers of Tamil Eelam Technological Innovations:
Purpose and Intended Mitigation of Government Countermeasures

Innovation	Purpose	Intended Mitigation of Government Countermeasures
Wax-coated wiring in explosive devices	Prevent emission of explosive vapors	Defeat detection by sniffer dogs
Airtight casing for explosive devices	Prevent emission of explosive vapors	Defeat detection by sniffer dogs
LED indicator lamps in bomb circuits	Verify "live" circuitry	
Secondary, tertiary detonation triggers in explosive devices	Provision of internal fail-safe mechanism	
"Explosive bra cup" design for suicide vest	Conceal explosive slabs	Defeat physical hand searches
Elongated fuel tank in vehicle bombs	Conceal explosive devices	Defeat detection by "dipper" probes
Chassis molded, mint-laced explosive devices	Conceal explosive charge and prevent emission of explosive vapors	Defeat casual visual inspections and detection by sniffer dogs
Hollowed out, shallow superstructure for suicide boats	Increase speed and reduce surface detection	Minimize radar cross-section
Penetration rods affixed to suicide boat prows	Amplify explosive force	Defeat hardened SLN superstructures
Minisubmarines for diver operatives	Covert de-bussing inside harbors	Defeat port harbor patrols
Prepaid SIM cards, single satellite signals for communication devices	Avail secure communication	Defeat government communication intercepts
Discursive writing, Slidex chart for coding communications	Avail secure communication	Defeat government counterintelligence

setbacks and exclusion zones can only be achieved by outfitting trucks with more destructive explosive payloads. Certainly the Tigers have the resources and skill to do this. However, large-scale bombings are likely to result in considerable numbers of auxiliary casualties, including civilians and foreign nationals, which would be sure to engender international outrage and effectively negate any claim to legitimacy the group might have. Prabhakaran, no doubt, fully appreciates that such fallout would severely jeopardize LTTE's overseas propaganda and attendant fund-raising efforts and, more seriously, signal it as a group of global concern. In the post-9/11 era, a designation of this sort carries particular significance, not least because it might place the Tigers within the scope of the U.S.-led war on terror.

The LTTE experience carries several important lessons germane to analyses and assessments of terrorist defensive technologies. First, the context in which counterstrategies take place is not static but continually evolving; as such, official efforts to mitigate these efforts need to be similarly dynamic and forward-looking. Second, the adoption of specific defensive measures is frequently tied to the group's wider operational agenda and should not, therefore, be viewed as a strict action-reaction dynamic (for example, the Tigers' decision not to employ more destructive truck bombs to overcome target hardening around strategically or symbolically important buildings). Third, and related to the above, the degree to which an organization seeks to protect a specific tactic's integrity generally reflects the extent to which the organization perceives the modality in terms of its overall offensive utility. Defensive technologies—and changes in emphasis therein—can, in other words, provide a potentially useful indicator of a group's evolving strategic and attack priorities, which can, in turn, help inform the manner by which a government shapes and allocates resources for its own mitigation policies. Finally, LTTE is evidence of the level of sophistication that an extremist entity can achieve in responding and defeating state-instituted measures against suicide terrorism. With growing fears of future martyr strikes taking place directly on U.S. soil, the lessons gleaned from the Sri Lankan theater will be of considerable help in informing U.S. law enforcement and intelligence of the type of "high-end" attacks that could occur in this country and how a group might seek to preserve the operational durability of these assaults.

CHAPTER FIVE

Provisional Irish Republican Army

Introduction

Over the course of its history, PIRA carried out a high-intensity campaign of terrorism, with the stated goals of bringing about unification of the six counties of Northern Ireland with the Irish Republic and the end of British involvement in Northern Ireland. Growing out of a much longer history of conflict, PIRA was born from a fracture in the Republican movement in 1969 when the group split off from what became known as the Official Irish Republican Army (Official IRA). Beyond the activities of Republican groups, the conflict also involves violence perpetrated by Loyalist organizations, which support continued English involvement in Northern Ireland. In addition to the political elements of the conflict, the division between Republicans and Loyalists is also largely a division between Catholics and Protestants. The religious dimension, though not an absolute division between the opposing sides, frequently made the activities of PIRA and its Loyalist opponents (e.g., the Ulster Defense Association and Ulster Volunteer Force) as much about brutal sectarianism as about the organizations' political aims (Coogan, 1993; Drake, 1991).

In 1997, PIRA officially agreed to a cease-fire as part of the ongoing peace process in Northern Ireland (*Jane's World Insurgency and Terrorism*, 2004), though the group has maintained its cohesion and carried on a variety of activities since that declaration.[1] Just as PIRA

---

[1] Personal interviews with law enforcement and government officials, Northern Ireland (May 2005).

split away from the Official IRA over political and strategic differences, PIRA's cease-fire resulted in the formation of splinter organizations, the Real IRA and Continuity IRA, that have continued to stage terrorist operations of a more limited scope since 1997. More recently, the group announced that it had decommissioned its arms as part of the peace process in Northern Ireland ("IRA 'Has Destroyed All Its Arms,'" 2005).

PIRA terrorism aimed at advancing its goals covered a range of activities, including assassination of specific individuals, attacks on people as members of specific classes of individuals (e.g., police officers and members of the security forces, government representatives, and sectarian targeting of Protestants); and attacks aimed at damaging specific physical targets and producing more generalized terror in the population (J. Bowyer Bell, 1993).[2] Although the majority of PIRA's operations were carried out within Northern Ireland, it also staged attacks and carried out other supporting activities on the British mainland, in the Republic of Ireland, and elsewhere, including continental Europe and the United States.

Compared with many terrorist groups, PIRA is an exceedingly sophisticated organization. Throughout its operational career, the group demonstrated significant technical acumen in manufacturing and improving offensive weapon systems, collecting intelligence, managing logistical operations, and training its members to carry out a variety of attack operations. The group maintained a cadre of technical experts and built a significant capacity to adapt and change over time, providing it with the resources and organizational ability to respond to the countermeasures fielded against it (Jackson, 2005). Government action against PIRA involved a range of organizations, including local law enforcement groups (in both investigative and intelligence roles), the British military, and British national intelligence organizations.

---

[2]   Personal interview with former law enforcement officers, Northern Ireland (May 2005). As part of its effort to maintain its image and legitimacy, PIRA sometimes provided warnings for some operations, creating an additional set of issues and dynamics with respect to defensive measures. The group's motives for providing warnings and, as a result, its intended level of accuracy and utility for preventing causalities were not always clear and, therefore, a matter of significant dispute.

The duration and scope of the conflict and the efforts against the organization mean the case includes examples from each of the five classes of defensive technologies defined in this book. Subsequent sections will examine PIRA's counterefforts aimed at technologies to

- acquire and manage information
- take preventive action
- deny access to potential targets
- respond to the effects of attacks
- investigate group members after operations.

Information Acquisition and Management

Because information on terrorist activities is one of the most potent weapons available to law enforcement and counterterrorist organizations, the area of surveillance was a primary field of technology and countertechnology competition in the Northern Ireland conflict.

Technologies Deployed

In its effort to gain advance warning of PIRA operations and information on its internal group activities, security forces fielded a variety of surveillance technologies and techniques against the group.[3] Surveillance modes applied include direct human surveillance, use of surveillance technologies, mechanisms for "community surveillance" (i.e., reports of suspicious behavior by private citizens), and the use of a number of information technology systems to process and apply the collected information.

---

[3] The primary government and security forces mode for gathering information on PIRA was direct infiltration of the group, either through recruiting current members as informers or placing agents in the group from the outside. Although infiltration is not technological and, therefore, not within the scope of this analysis, the absence of discussion of this topic should not be interpreted as a judgment of its value or applicability in an overall approach for combating terrorism. In fact, much of the information in the following discussion on technologies and PIRA's countertechnology strategies is derived from the first-person accounts and other reports provided by these individuals.

Direct human surveillance. Within Northern Ireland in particular, surveillance activities by law enforcement officers and military personnel were a key element of surveillance operations (Barzilay, 1981). Vehicle checkpoints, set up frequently to protect areas from attack, provided the opportunity to collect broad baseline information on the population and their movements. "Stop and search" operations against individuals and general police observation of individuals on the street in search of known or suspected PIRA members provided similar opportunities (Dewar, 1985; Dillon, 1990). Activities such as neighborhood surveys and intrusive searches of homes and commercial buildings were used to build a data set on the nature of the urban environment in which PIRA was operating (Dewar, 1985).

Reflecting the intensity of the ongoing conflict in Northern Ireland, permanent observation posts were built in towers or atop tall buildings that provided a stationary and overt platform for monitoring the city.[4] In the posts "such as the one established on top of Divis Flats in West Belfast, there [were] several observers continuously scanning the streets of the Lower Falls area using high-powered binoculars and, at night, infrared sights" (Dillon, 1990, p. 409). Covert surveillance posts were used as well by placing soldiers or others in concealed locations in areas of interest (or near specific locations such as suspected PIRA safe houses or arms dumps). These included infiltration of observers into attics or derelict buildings where they could observe traffic and passersby and record data over a period of several days (Dillon, 1990).

Surveillance technologies. During the approximately three decades of the conflict, a wide variety of surveillance technologies were developed and deployed as part of the security effort (Geraghty, 2000). Like overt observation posts, significant numbers of visible surveillance devices, such as CCTV and other systems, were installed in areas of PIRA operations (Coaffee, 2004).[5] As technologies advanced, more and more processing and analysis capabilities

---

[4] Personal interviews with law enforcement and former law enforcement members, Northern Ireland (May 2005).

[5] Personal interviews with law enforcement and former law enforcement members, Northern Ireland (May 2005).

were integrated into these systems. As described by Geraghty (2000, p. 163), "Surveillance cameras around sensitive areas such as the City of London, linked to computers which will automatically identify suspect vehicles within four seconds, evolved into computerized digital maps of human faces."

A variety of devices was integrated into mobile platforms such as the aircraft and helicopters that frequently patrolled areas in Northern Ireland. Technologies carried by these vehicles included sophisticated photographic devices, live-feed television cameras, and detection systems such as infrared thermal sensors that could detect soil that had been recently disturbed to assist in locating PIRA land mines, detonation wires, or underground arms dumps or facilities (Barzilay, 1973; Dillon, 1990; Geraghty, 2000; Urban, 1992). Technologies aimed specifically at detecting PIRA weapons were also used by security forces. To counter the broad use of explosives in the conflict, soldiers and law enforcement officers used trained dogs, technological "sniffers" for explosives, and technologies such as mobile x-ray platforms to screen for bombs in vehicles (Barzilay, 1973; Ryder, 1997; Styles, 1975).

Although information on many of the covert surveillance technologies that were applied during the conflict is still limited in the open literature, it is clear that an extremely wide range of such devices was used to collect information on PIRA activities. Listening devices, phone taps, hidden cameras, motion detectors, and technologies that allowed interception of communication traffic played critical roles (Adams, Morgan, and Bambridge, 1988; Dillon, 1990; Geraghty, 2000; Taylor, 2001). A variety of devices was reportedly used that could be deployed in areas of interest—from zones where PIRA operatives moved across the border between Northern Ireland and the Republic of Ireland to underground tunnels where terrorist operations were suspected (Dillon, 1990; Geraghty, 2000).

Devices attached to suspect vehicles that transmitted a signal to allow tracking of the sensor's position were also used (Dillon, 1990; McGartland, 1997). Some were even implanted in discovered PIRA bombs and weapons (known as "jarking" the weapons) to track their movement from weapon dumps to other locations used by the terrorists (Dillon, 1990; Geraghty, 2000; McGartland, 1997; Urban, 1992).

Community surveillance. The level of intimidation of the local community by PIRA and other paramilitary organizations made most citizens hesitant to provide any information about the groups' activities to the police or the military. To provide a route through which individuals could do so without exposing themselves to the same level of risk—for example, being seen entering a police station—the security forces set up the Confidential Telephone system so individuals could pass on information anonymously (Barzilay, 1973; Ryder, 1997). Although the system produced some problems for the police (discussed below), it also produced some valuable intelligence (Ryder, 1997).

Information management systems. Because of their extensive surveillance capabilities, the security forces in Northern Ireland and the British mainland had large amounts of intelligence on PIRA members and operations. From relatively straightforward activities such as monitoring Republican gatherings and public demonstrations, databases were created of potential group members and supporters; from more sophisticated operations and systems, specific data were collected on individuals and their activities.

Sophisticated information management systems were needed to use such a large volume of information effectively. Starting from basic banks of card files and listings of photographs of potential PIRA terrorists or sympathizers (Barker, 2004; Urban, 1992), these tools evolved into complex databases and computerized information management systems as the conflict progressed. Reports of security force activities indicate that there were computer systems focused on collecting data on vehicles in Northern Ireland (code-named Vengeful) and a similar system for data on individuals (code-named Crucible). Crucible has been described as holding personal data, maps, photos, and information on individuals' locations, family connections, and past activities (Barker, 2004; Barzilay, 1981; Geraghty, 2000; Urban, 1992). As processing capabilities increased, additional knowledge-based capabilities were reportedly integrated into the computer systems to improve data analysis and pattern recognition (Geraghty, 2000). This helped perfect the application of techniques such as traffic analysis and network analyses of groups (van Meter, 2002) and also enabled the security forces to recognize even small changes in suspects' behavior. For example, if the systems detected specific PIRA suspects not appearing where they were expected or dropping out of sight, attention was then focused on locating those individuals and determining the reasons for the anomaly (Canadian Security Intelligence Service, 1994; Dillon, 1990).

Countertechnology Responses by PIRA

In an effort to blunt the impact of security forces' information-gathering activities and the corresponding disruption of its activities, PIRA implemented a range of counterefforts. PIRA developed approaches to *evade* the surveillance methods, to *conceal* the signatures or features the surveillance methods were designed to detect, and to directly *attack* the technologies themselves.

Evasion. The most basic responses instituted by PIRA against the surveillance activities of the security forces were simply efforts to avoid areas they believed were monitored, or technologies they thought were easily penetrated by the security forces (such as communication). Because of the general divisions that exist within Northern Ireland— where there are areas widely known to be dominated by Republican supporters—the first component of these strategies was to operate against this broadly known pattern of behavior (Dillon, 1990). PIRA also displaced their attack activities from areas of intense vigilance and security to those where it was lower (O'Callaghan, 1999). This included an effort to have PIRA cells operate outside their home areas in an attempt to reduce the chances they would be recognized, though the challenges this posed for operations made it difficult to implement (Coogan, 1993). It is also observable in locations of attacks in London after installation of the surveillance-heavy "Ring of Steel" to protect the financial district—later operations were staged outside the coverage of the system[6]—and in PIRA carrying out attacks outside Northern Ireland in general as security was increased (Drake, 1991).

PIRA also avoided technologies it believed were readily monitored, such as the telephone. For example, "in the Sinn Fein offices in Falls Road, there hangs a warning sign which says: 'This phone is bugged'" (Adams, Morgan, and Bambridge, 1988, pp. 4–5; also discussed in

---

[6] Personal interview with British law enforcement official, California (February 2005).

Barzilay, 1975). Pressures placed on the use of traditional communication imposed considerable operational costs on PIRA and forced it to adopt alternative communication modes. Some were of limited utility; for example, "Coded signals in newspaper small-ads lacked the flexibility of two way conversation" (Geraghty, 2000, p. 154). PIRA cycled through alternative technologies such as pagers and email (Geraghty, 2000) until concerns about their being compromised arose as well. As has been observed with other terrorist organizations, the group has recently made use of mobile phones in concert with strong security measures, including frequently replacing its phones and fully neutralizing them prior to operations—either leaving them behind or ensuring that their batteries have been removed.[7]

Due to suspicions about communication monitoring, PIRA frequently carried out much of its business face-to-face and did so in a way that limited the ability of other surveillance technologies to monitor the conversations. The group reportedly carried out "walk-and-talk" briefings during which individuals conversing moved from place to place outdoors in an effort to conceal their discussion and avoid fixed site surveillance assets.[8] Descriptions of behavior reported by infiltrators inside PIRA indicated that certain conversations were held outside buildings to avoid any listening devices that might have been present inside; other interchanges where visual clues—such as examining maps—might provide insights to security force observers were held only inside (McGartland, 1997). PIRA was similarly suspect of using any single location regularly for sensitive communication[9] or hiding weapons (Barzilay, 1975), and members "changed their cars" regularly—by stealing new ones—to defeat attempts to plant listening or tracking devices in them (Geraghty, 2000, p. 147).

PIRA evasion efforts were supported, where possible, by detailed studies of the limits of the surveillance technologies. There are reports of PIRA conducting extensive "dead ground studies" to determine the visual ranges of specific observation posts (Harnden, 2000, p. 259), as

---

[7]   Personal interviews with law enforcement, Northern Ireland (March 2004, May 2005).

[8]   Personal interviews with law enforcement, Northern Ireland (May 2005).

[9]   Personal interviews with law enforcement, Northern Ireland (May 2005).

well as doing surveys of CCTV coverage in areas before they carried out operations.[10] When walking through areas that the group knew would be monitored from the air using technologies such as infrared, its members learned to walk near high hedges or animal paths that would hide evidence of their footprints from the sensors above (Dillon, 1990). They also became very sensitive to specific weather conditions that were problematic for different types of sensing devices—such as high winds, fog, and rain—and used them to provide additional cover from observation (Dillon, 1990; Harnden, 2000).

Concealment. When they could not evade surveillance technologies, PIRA members made efforts to conceal themselves from them— to obscure whatever behavior, features, or signature the technology or surveillance effort was designed to detect. This included basic behavioral activities such as learning how to "look natural" when they knew they were being observed[11] and to ensure that individuals' behavior did not betray their connections to the Republican movement or PIRA:

> Never talk loosely and be constantly on your guard and on the look-out. . . . Keep well away from Republican marches and protests, so that you don't become known to security forces. . . . If you go to a pub after the job, never show any give-away signs. Don't be getting the staff to switch on the TV or the radio so that you can listen to reports about the job. Just be cool, discreet and professional. (Gilmour, 1998, pp. 99–100)

Similarly, basic behavioral changes have been adopted, such as the use of disguise (O'Ballance, 1981), PIRA members' emblematic use of balaclava face masks, masking of license plates on cars used in operations (Ryder, 1997), and wearing less obvious items such as baseball caps, which, depending on the design of surveillance or CCTV systems, may be enough to obscure the wearer's identity (McGartland, 1997).[12]

---

[10] Personal interviews with law enforcement, Northern Ireland (May 2005).

[11] Personal interviews with law enforcement, Northern Ireland (May 2005).

[12] Personal interview with former law enforcement official, England (May 2005).

PIRA developed the capability to forge identity documents to provide group members apparently established identities that appeared clean to the security services.[13] The group applied similar deception techniques to fool the security services' vehicle identification systems:

> [PIRA] operatives toured the streets of prosperous areas, whose inhabitants would be listed in VENGEFUL as being of no interest, and took the precise details of cars. They would then find a similar model, change its number plates [to match the "clean" automobile] and ensure that it was identical to the first, even down to stickers in the window. In this way a soldier or police officer checking the number by computer would assume the car belonged to a respectable suburbanite. (Urban, 1992, p. 115)

In some individual cases, the group went to even more significant lengths to "clear" individuals from security forces' surveillance lists; in at least one instance, the group reportedly took advantage of a bombing whose victims could not be identified and announced the death of a prominent PIRA member, so he could be "resurrected" later when the security forces had stopped paying attention to him (Coogan, 1993). A more basic response to security forces' surveillance of individuals was a trend in PIRA operations to use "unknowns"—members who were new to the group and therefore unlikely to be singled out for attention (Canadian Security Intelligence Service, 1994; McGartland, 1997).

The group also made procedural and technical changes in an effort to conceal the behavioral and other signatures of its terrorist activities. When the group was forced to increase the destructive power of its bombs—therefore making them more and more unwieldy and requiring more members to transport—it transitioned to the use of bombs built into vehicles to limit the number of individuals required for operations and, therefore, the obviousness of the behavior (Drake,

---

[13] Personal interview with law enforcement officials, Northern Ireland (May 2005). In some cases, the elaborate nature of the cover identities for individuals operating on the British mainland—PIRA had provided its operatives with more corroborating documents than a "normal person" would likely have—was a signal to the security services (personal interview with law enforcement official, England, May 2005).

1991). When explosives technologies like Semtex[14] became available that provide more explosive power than homemade materials, PIRA transitioned to smaller bombs that were easier to conceal and moved away from explosives based on ingredients (such as fuel oil or nitrated-benzenes) that had distinctive odors and were therefore more readily detectable.[15] It also took steps to conceal the presence of bombs through techniques such as wrapping the explosives in many layers of cellophane to limit the ability of technologies to detect them (Holland and Phoenix, 1996). The group changed the construction of underground arms caches and logistics facilities to reduce their profile for aerial infrared and other detection devices (Dillon, 1990; Geraghty, 2000; Horgan and Taylor, 1997; McGartland, 1997).

PIRA also took steps to reduce the time that operatives were actually in possession of weapons during operations to the absolute minimum—since an individual carrying a weapon was easy to link with violent activity. This included, for example, elaborate chains of operatives and supporters delivering the rifle to a sniper and, immediately after the shot was taken, removing it to a local arms hide (Marques, 2003). PIRA addressed a similar problem with grenade launchers via another strategy: Because the group had only a small number of commercial grenade launchers, after using them, the terrorists had to carry the launcher away from the scene. This made them very obvious and easy to apprehend. To address this, the group turned to manufacturing its own launchers so the tubes became "disposable," and the operative could simply discard it at the scene of the attack and make his escape (Geraghty, 2000).[16]

Attack. When PIRA has had the opportunity, it has also attacked surveillance technologies directly. These attacks included fielding countersurveillance assets and procedures aimed at detecting, uncovering, and confusing human surveillance teams. A variety of approaches was used, including stationing its own surveillance teams near its weapon

---

[14] Semtex® is a registered trademark of Explosia, a.s.

[15] Personal interviews with law enforcement technical experts, Northern Ireland (May 2005).

[16] Personal interviews with law enforcement officials, England (March 2004).

caches to detect security force surveillance efforts (Morris Tribunal, 2004); posting lookouts to identify and call out security force watchers (Urban, 1992); using "challenge-response studies" to see how security forces would tip their hand when known, high-profile PIRA members appeared in an area;[17] and applying operational procedures designed to convince surveillance teams they had been "made" so they would break off surveillance. "As one officer explained: 'IRA suspects sometimes made a meal of countersurveillance. I remember tailing a Provo[18] car in London. It roared several times round the same roundabout while the driver wound down his window and lifted two fingers.' In fact, the same terrorist did this routinely, as prescribed by his trainers, with no knowledge of whether or not he was being followed" (Geraghty, 2000, p. 150).

PIRA also fielded efforts to detect and eliminate surveillance devices deployed by the security forces. The group reportedly developed or procured equipment to assist in detecting listening devices (Dillon, 1990)[19] or countering their effectiveness (Ryder, 1997) and coupled use of such equipment with rigorous processes of examining weapons (Dillon, 1990) and "carrying out minute examination of vehicles and premises to ensure [that] no listening or tracing devices [were] installed" (Ryder, 1997, p. 351). More basic approaches such as simply pulling telephone sockets out of the walls of safe houses (to defeat any bugs that relied on them) were also used (McGartland, 1997). The group also reportedly sought to counter the placement of transmitters in weapons through more rigorous control of who had knowledge of weapon dump locations—thereby making the discovery of a jarked weapon an opportunity for the group to root out potential informers (McGartland, 1997; Urban, 1992). The effectiveness of the countermeasures adopted by the group reportedly meant "the police had virtu-

---

[17] Personal interview with law enforcement, Northern Ireland (May 2005).

[18] *Provo* is a common nickname for a PIRA member.

[19] However, some individuals questioned how much of PIRA's ability to detect surveillance devices was simply rigorous procedures coupled with good luck rather than the adoption of technical countermeasures (personal interview with government security official, Northern Ireland, May 2005).

ally to abandon all technical surveillance operations for a considerable time" (Ryder, 1997, p. 351).[20] In response to the use of metal detectors as part of the military's effort to detect PIRA bombs, the group allegedly developed specific electromagnetic traps that would detonate the device when exposed to the signals transmitted by the detectors (Geraghty, 2000).

To attack the information systems that the security forces relied on, PIRA applied a variety of hoax and deception techniques. The group used hoax "informant" calls identifying innocent people as members of the group (Geraghty, 2000; Marine Corps Intelligence Activity, 1999). PIRA also took more direct approaches to remove data from the hands of the security forces: "As recent as 2002 . . . a pair of guerrillas gained access to a police barracks using fake or stolen identity cards . . . , overpowered a lone guard inside and stole several police files on IRA members" (Marques, 2003, p. 29).

PIRA also attacked a community tip line operated by the police by turning it into a vehicle of attack on the security forces themselves: "Although [the Confidential Telephone] produced some useful information—the [police] said that 500 calls had been of value—it was also used by the terrorists to lure the security forces into ambushes or booby-traps" (Ryder, 1997, p. 124; MacStiofáin, 1975, p. 331).[21] Sean MacStiofáin, one of the early leaders of PIRA, described the effects of the campaign: "This was one of several ways in which the informer-phones were played back against them, with the result that many of the military came to mistrust what they had thought was a foolproof way of getting contact intelligence" (MacStiofáin, 1975, p. 331). Such hoax calls were also made to the line that citizens called to request police assistance—raising the potential that any call for help from a citizen might be a trap; this behavior by PIRA resulted in the police being

---

[20] One reason that has been suggested for PIRA's strenuous pursuit of surface-to-air missiles focuses on its role in security forces' surveillance activities: If the group could attack those platforms directly, it could take away a key information-gathering resource (see Dillon, 1990). Fortunately, the group was largely unsuccessful in this effort.

[21] Personal interview with law enforcement officer, Northern Ireland (May 2005).

forced to institute call-back procedures to callers to reduce the danger to responding officers.[22]

Preventive Action

Because a group's capability level affects the level of threat it poses, degrading PIRA's overall organizational capabilities was a key element of security forces' effort to combat the group's terrorist activities.

Technologies Deployed

PIRA's weapon technologies were relatively advanced for a terrorist organization. As a result, myriad defensive technologies were designed to directly undermine the value and effectiveness of the group's weapons. These included development of a wide range of electronic tools for defeating the ways PIRA detonated its IEDs and triggered other weapons. Early versions of the devices included sweep transmitters aimed at predetonating radio-controlled bombs (Barzilay, 1975; MacStiofáin, 1975).[23] Later generations focused on the development of sophisticated jamming equipment to interfere with or suppress the detonation systems (Harnden, 2000; Urban, 1992).[24] In some cases, specific countermeasures were installed for particular weapons, such as antimissile systems reportedly put on helicopters operating in Northern Ireland to counter the potential threat from surface-to-air missiles (Harnden, 2000).

Efforts were also undertaken to limit PIRA's offensive capability by reducing access to specific types of weapons and the components for manufacturing them. Significant monitoring and interdiction efforts were focused on PIRA attempts to import weapons from outside of the Republic and Northern Ireland, resulting in several high-profile seizures of major shipments (O'Callaghan, 1999). Efforts were also put in place to limit the flow of explosives into the province (Hamill,

---

[22] Personal interview with law enforcement officer, Northern Ireland (May 2005).

[23] Such an approach has obvious limitations, particularly in an urban environment.

[24] Personal interview with former security forces member, England (March 2004).

1985), reduce the chance of commercial explosives being diverted for terrorist purposes, and reduce the availability of ingredients for improvised explosives by controlling sales of some materials and reformulating others (Barzilay, 1975; Commission on Physical Sciences, 1988; Foulger and Hubbard, 1996).[25]

Beyond PIRA's offensive capabilities, the security forces also made significant efforts to undermine other elements of the group's capabilities. For example, police and security forces frequently seized or rendered nonfunctional PIRA weapon stocks that had been identified through tips or surveillance (Barker, 2004; Geraghty, 2000; Gilmour, 1998; McGartland, 1997).[26] Similarly, efforts to curtail PIRA's access to financial resources did not depend on specific technologies. Because much of the group's financing was generated through illegal activities such as extortion, smuggling, fraud, drinking clubs, and other enterprises (Harnden, 2000; Horgan and Taylor, 1999, 2003; McGartland, 1997; O'Callaghan, 1999), efforts to curtail money flows relied heavily on traditional police methods for fighting organized crime (Horgan and Taylor, 2003). Such efforts were and, in fact, remain important components of the overall struggle against the group.[27] However, since they usually do not depend on specific defensive technologies, they are not as germane to the current analysis of PIRA countertechnology activities.

Countertechnology Responses by PIRA

In all areas affecting its group capabilities, PIRA made vigorous efforts to counter the effects of defensive technologies. Efforts by security forces to jam the electronic signals that the group used to detonate its explosives led to a stepwise technology race between the two sides. Back-and-forth modifications included changing the frequencies used to trigger the bombs, changes in the coding of the signals, and

---

[25] Personal interview with law enforcement technical experts, Northern Ireland (May 2005).

[26] Personal interviews with former security forces member, England (March 2004) and law enforcement, Northern Ireland (March 2004).

[27] Personal interview with law enforcement, Northern Ireland (May 2005).

even incorporating completely different technologies such as photographic flash units and radar detectors to trigger detonation (Geraghty, 2000; Harnden, 2000; Urban, 1992; Jackson, 2005). The group also responded by revisiting older methods of detonating devices—such as basic command wires—that provided alternatives to the technologies targeted by the security countermeasures (Urban, 1992).[28] PIRA also substituted other attack modes once explosives became harder to use. Its transition to sniper tactics has been described as "a response to the Army's success at jamming radio-controlled devices" (Harnden, 2000, p. 406). A former PIRA leader also alleged that the group developed ways to "home in" on the signals the countermeasure systems emitted and use the information to target the security forces (MacStiofáin, 1975, p. 235).[29]

PIRA's countermeasures to the efforts of the security forces to restrict the flow of explosives and their ingredients were largely attempts to innovate around the constraints. Controls on commercial explosives made them less readily available, leading to an increase in the use of homemade explosives (Barzilay, 1975; Commission on Physical Sciences, 1988). Controls on and modifications to specific bomb-making ingredients similarly increased the difficulty for the bomb makers. For example, although it has been reported that the conversion of Irish agriculture from pure ammonium nitrate–based fertilizers to calcium ammonium nitrate (a formulation less readily made into bombs) resulted in fewer bombing incidents, PIRA bomb makers found ways to convert the material into a form that was usable for their purposes (Commission on Physical Sciences, 1988). PIRA also substituted other weapons, such as blast incendiaries, when explosives were unavailable (Hamill, 1985; Ryder, 1997). The availability of such alternatives reduced the impact of the control efforts on the capabilities of the group.

---

[28] Personal interview with former security forces member, England (March 2004).

[29] Other sources suggest that the security forces maintained this capability to triangulate the locations of PIRA operatives based on the radio signals intended to detonate their bombs (see Barzilay, 1981).

Denial

Given the scope and duration of PIRA's terrorism campaign, a wide variety of efforts to harden or protect potential targets was instituted in both Northern Ireland and on the British mainland. Technologies and approaches were shaped to protect the full range of PIRA's targets, from individuals—particularly members of the security forces—to locations and areas that were frequent targets of attack.

Technologies Deployed

In response to PIRA's violent activities targeting individuals, efforts were made to guard the identities of potential targets—such as members of security force organizations—and to protect them to the extent possible (Holland and Phoenix, 1996; McGartland, 1997). Technologies fielded to do so included traditional protective devices such as bullet-resistant body armor ("Five Days in an IRA Training Camp," 1983; Ryder, 1997), as well as such approaches as heavy armoring of security force vehicles to withstand most firearms and explosives attacks (Barker, 2004; Ryder, 1991, 1997).[30] In some areas, there were shifts in transportation used by the security forces—to aerial modes, predominantly helicopters—in an effort to put one of the organization's primary targets "out of reach" of most of its available weaponry (Harnden, 2000, p. 19).

Frequent targets such as police stations and other government buildings were gradually fortified against PIRA's methods of attack (Barker, 2004; Harnden, 2000; Murphy, 2005; Ryder, 1997).[31] To deter attacks being staged in the first place, security patrols were also deployed around potential targets so attack operations could not be initiated (Urban, 1992).

Areas of major targeting, such as the Belfast city center (Ryder, 1997) and the Square Mile—the heart of the London financial district—were hardened extensively against attack inside security mea-

---

[30] Personal interview with former security forces member, England (May 2005).

[31] Personal interviews with former security forces members, England (March 2004) and with law enforcement member technical expert, Northern Ireland (May 2005).

sures that came to be known colloquially as "Rings of Steel" (Coaffee, 2004). The areas were closed to most vehicular traffic, and massive surveillance and guard nets were put in place to monitor and search pedestrians and those vehicles that were admitted. Gates were installed to close off the center of Belfast at night, and parking unattended vehicles was banned during the day to prevent the planting of car bombs (Ryder, 1991).[32]

Countertechnology Responses by PIRA

In response to the significant hardening efforts, PIRA adopted strategies to reduce the effectiveness of the security measures and reconstitute the group's ability to inflict harm. PIRA's specific approaches fall into three general strategies: escalating to larger weapons, adopting new weapons, and attempting to avoid the defensive measures.

Escalating. Simply increasing the size and scale of weapons applied—the larger bomb, the heavier mortar—can frequently be successful in overwhelming a strengthened defense (Murphy, 2005; Ryder, 1997). As bollards and barriers were installed to keep vehicle bombs back from major targets, the result was larger and larger truck bombs that produced escalating amounts of collateral damage to adjacent structures.[33] A prominent example was PIRA's attack on the police forensics facility where a multithousand-pound bomb was used to compensate for the facility's setback from the road. The device resulted in significant damage to several hundred homes in the surrounding area (Ryder, 1997).[34] In some cases, in addition to escalating the size of the devices, the group placed them outside the hardened defenses, simply accepting that their effectiveness would be reduced; for example, responding to the perimeters and cordons designed to keep vehicle bombs outside of central Belfast, PIRA sometimes just tried to get as close to the denied area as possible and set off the devices there.[35]

---

[32] Personal interview with law enforcement officer, Northern Ireland (May 2005).

[33] Personal interview with former law enforcement officer, Northern Ireland (May 2005).

[34] Personal interview with former security forces member, England (May 2005).

[35] Personal interviews with law enforcement officer, Northern Ireland (May 2005) and with former security forces member, England (May 2005).

Adopting new weapons. In a number of cases, PIRA developed or adopted novel weapons or tactics in response to defensive hardening. The group's use of mortar bombs is an example of this behavior. Denied much of its ability to attack sites such as police stations by the cordons and walls placed to protect them (Holland and Phoenix, 1996), PIRA sought to attack them from a direction lacking armor—above the buildings through the use of mortars (Coaffee, 2004).[36] In response to the gradual strengthening of the armor on security forces' vehicles, PIRA developed its own armor-piercing weapons—based on explosives designed with shaped charges for increased penetration (Urban, 1992).[37] The devices, delivered through a variety of methods (Geraghty, 2000; Jackson, 2005), made it possible for PIRA to penetrate significantly armored targets.[38] The group also adopted the use of explosives coupled with large amounts of flammable materials—gasoline tanker trucks—in an effort to overcome defensive measures (Patrick, 1981).

Confronted by protective measures, PIRA also pursued new weapons that better matched the group's needs as its circumstances changed. In response to security forces' extensive use of aircraft for transport, the group strenuously pursued weapons that would allow it to attack aerial targets ("IRA Interview," 1981; Harnden, 2000; Urban, 1992). In urban areas, the group shifted from major bombs—they were effectively prevented entry through checkpoints and roadblocks—to small incendiary devices that could be more easily smuggled.[39] The group also made lower-level shifts in the way it constructed its weapons to make avoiding security measures easier. Rather than building an entire bomb and delivering it whole to a target, the group smuggled

---

[36] Personal interview with law enforcement technical expert, Northern Ireland (May 2005). Reportedly, PIRA attempted an alternative mode of delivering such weapons from above—dropping explosive devices from a hijacked helicopter (Bell, 1998). The experiment was not successful, and mortars remained the group's primary mode for attempting such an attack.

[37] Personal interview with former security forces member, England (May 2005).

[38] Personal interview with law enforcement technical expert, Northern Ireland (May 2005).

[39] Personal interview with law enforcement officer, Northern Ireland (May 2005).

small amounts of explosives through security cordons over longer periods and assembled the bomb to or near the desired target (Barzilay, 1975). In addition, the group also adopted the use of timing devices that enabled delays over many days or weeks—such as that used in the attack on Prime Minister Margaret Thatcher at Brighton—that allowed planting of the device before a security cordon was even put in place (Geraghty, 2000).

Avoiding defensive measures. Rather than making major shifts in weaponry in response to defensive measures, sometimes PIRA simply tried to avoid them. Some defenses resulted in operational shifts by the group to new tactics—for example, the adoption of sniping to target security forces when defenses made it more difficult to use explosives (Harnden, 2000). The group also adjusted its application of particular tactics in response, such as shifting the points at which group members were instructed to aim their weapons based on the adoption of body armor by security forces and improvements in the armoring of vehicles ("Five Days in an IRA Training Camp," 1983). As has been demonstrated for other terrorist organizations (Enders and Sandler, 2004), in some cases, PIRA simply relocated its violent activities from defended to less-defended areas and targets. This included shifts in the areas of Northern Ireland that were targeted (O'Callaghan, 1999), movement of the locations of attacks on the British mainland (Bell, 1998; Jenkins and Gersten, 2001), and increasing use of hoaxes as compared with actual bombings (Jenkins and Gersten, 2001).

The group sought to use deception to allow it to penetrate defenses without impacting its operations. In a strategy now seen in several terrorist groups, PIRA recruited women to carry weapons through cordons, because they were less thoroughly searched by predominantly male security forces (Barzilay, 1973). Later, a high-profile PIRA response to the effectiveness of vehicle cordons was the use of so-called proxy bombs: compelling innocent individuals to transport bombs to their targets through violent threats to them, their property, or loved ones (Barzilay, 1975; Coogan, 1993; Drake, 1991; McGartland, 1997; O'Ballance, 1981). This strategy defeated a number of security measures such as the use of authorization or "admittance passes" for access control, security force databases on suspect vehicles, and monitoring of suspected PIRA members or sympathizers.[40]

Response

Although many of the effects of a terrorist attack occur immediately—damage caused by a bomb, people killed in an armed assault—others can be reduced or eliminated through rapid and effective response action. Responses to attacks that are in progress, yet not fully realized, potentially can fully defeat their attempt to cause harm. As a result, building capabilities among the security forces and response organizations in Northern Ireland and the British mainland was a key element of the overall effort against PIRA and its terrorist campaign.

Technologies Deployed

Over the course of the conflict, security forces and response organizations in Northern Ireland became adept at responding to terrorist activities to reduce their overall impact. Some of these capabilities were distant from PIRA's activities and were therefore not something that the group would seek to counter; for example, hospitals in Belfast developed internationally recognized expertise in addressing the types of traumatic injuries produced by terrorist activities as well as those from the "punishment" beatings and woundings that made up PIRA's "community policing" activities (Stevenson, 1994).

Others, however, presented a more serious challenge to the capabilities of the organization. These include the ability to respond to the immediate outcome of incidents and, most importantly, the ability of explosives response teams to prevent the detonation of PIRA bombs. This capability was supported by training the public to recognize and report unattended items and, in the event of a telephoned bomb threat, to collect needed information to support rapid and effective responses

---

[40] Personal interviews with former and current law enforcement officers, Northern Ireland (May 2005); with law enforcement officers, England (March 2004); and with law enforcement official, Northern Ireland (March 2004).

(Jenkins and Gersten, 2001). The bomb disposal teams, armed with a wide variety of tools and devices for dismantling and "rendering safe" planted bombs (e.g., Barzilay, 1978), could frequently dismantle or disrupt bombs in only minutes (Patrick, 1981; Styles, 1975).[41]

Countertechnology Responses by PIRA

Because the security forces were one of PIRA's primary targets, it is frequently difficult to judge whether specific steps taken by PIRA were aimed at countering security forces' response capabilities or simply extensions of the group's effort to kill and injure soldiers, police officers, and bomb disposal officers (Styles, 1975). Irrespective of PIRA's fundamental intent, which likely included both, the group's actions did hamper the security forces' ability to respond to terrorist incidents.

Responding to the increasing skill of bomb disposal teams, PIRA's most basic response was to decrease the warning times it provided for bombs[42]—and the delays it set on the timers—to lessen the window of time in which bomb technicians had to work (Barzilay, 1981; Collins and McGovern, 1998; Ryder, 1997; Styles, 1975).[43] PIRA also sought to saturate response capabilities. For example, at one point, there were three explosives disposal teams in Belfast. Use of more than three bombs at one time or, more frequently, a combination of several bombs and many hoax devices or calls could overwhelm the ability to respond no matter how fast the teams operated.[44] Even if it was known that many were hoaxes, they had to be treated as potentially real and therefore the use of this tactic significantly hurt response capacity and increased the level of "chaos" inherent in any such response operation (Barker, 2004, p. 73).

PIRA also sought to injure and kill those who responded to attacks. In addition to using fake calls for assistance or crimes to bait

---

[41] Personal interview with former security forces member, England (May 2005).

[42] PIRA frequently called in warnings before bombing operations to theoretically provide sufficient time for security forces to clear civilians out of the area, but not enough time to defuse the explosive device.

[43] Personal interview with former security forces member, England (May 2005).

[44] Personal interview with former security forces member, England (May 2005).

them into ambushes (discussed above), PIRA incorporated a variety of booby traps into its explosive devices to hinder defusing and create the potential to harm responding officers. These booby traps were triggered by movement, light activation, and other strategies (Barzilay, 1973, 1978, 1981; Collins and McGovern, 1998; Geraghty, 2000; Hamill, 1985; O'Ballance, 1981).[45] PIRA also planted secondary devices targeting responders at scenes (Harnden, 2000; McGartland, 1997), the placement of which was frequently informed by studying response operations at previous incidents or hoaxes.[46] At least one first-person report by an infiltrator of PIRA indicates such bombs were sometimes specifically designed to prevent responders from assisting in the aftermath of an attack. A PIRA commander is quoted as directing his cell to "place the bomb by the exit to stop the emergency services, the peelers [police] and the ambulance men getting into the bar to attend to the dying and the injured" (McGartland, 1997, p. 263).[47]

Investigation

The operational focus on arresting and prosecuting individuals involved in terrorism placed a heavy focus on technologies and approaches for investigating terrorist incidents and for identifying those involved. Often drawing on intelligence tips from individuals or sources that had to remain secret, security organizations frequently had difficulties converting "intelligence into evidence" for prosecutions (Ryder, 1997,

---

[45] It has also been suggested that the incorporation of motion-sensitive antihandling devices were targeted not only at the security forces but also at members of the public who might pick up and move PIRA bombs to safe areas away from where they were planted (Barzilay, 1975).

[46] Personal interview with law enforcement officers, Northern Ireland (May 2005).

[47] PIRA also reportedly took advantage of jurisdictional divisions to limit the ability of the police to effectively respond to its operations. Some activities—such as interrogating suspected informers—might be carried out in the Republic of Ireland and any corpses that resulted dumped in the North. As a result, "The forensics are in the South and then the people who have to investigate that murder are the RUC [the police in Northern Ireland]. It [disrupts] their investigation" (Harnden, 2000).

p. 147).[48] That operational reality, coupled with the effectiveness of intimidation of potential witnesses by the paramilitary groups, put a premium on the use of forensic science as part of the overall effort to combat PIRA terrorism (Ryder, 1997).

Technologies Deployed

To investigate terrorist activities, the full range of forensic science techniques was deployed by the security and law enforcement organizations operating in Northern Ireland and on the British mainland. Described by Tony Geraghty in his book (2000) as "the forensic battleground," the techniques included DNA analysis, ballistics, handwriting and document analysis, examination of the military equipment produced by PIRA to support its operations, chemical analysis of explosive and weapon residues, hair analysis, fingerprint and other biometric analysis, and trace evidence analysis such as hair and fiber examination (Geraghty, 2000; Ryder, 1997).[49]

Countertechnology Responses by PIRA

PIRA leadership recognized the seriousness of the threat that forensic science posed to the group. Most of the group's responses focused on changes in operational practices to minimize the evidence left after an operation. These included selecting clothing that would not leave behind incriminating fibers and ensuring that members wore outer garments that could be rapidly washed or destroyed to eliminate traces of explosive or gunpowder residue (Geraghty, 2000; McGartland, 1997). Recently, there have been reports of operatives wearing forensic suits—the disposable garments used by forensic investigators to avoid contamination of a crime scene—during operations (Independent Monitoring Commission, 2004). In addition to washing or destroying clothing, members were also instructed to bathe as soon as possible

---

[48] Personal interview with law enforcement officers, Northern Ireland (May 2005).

[49] Descriptions of forensic examinations of PIRA attack scenes—such as the high-profile bombing of a fish shop on the Shankill Road—or PIRA armed attack operations provide specific data on the scale of forensic operations required after such an incident, the range of techniques applied, and some of the unique complications involved in investigating the scene of terrorist incidents (see McCorkell and Griffin, 1998; Quinn, 1998).

after an operation to remove any incriminating residues from their skin (Urban, 1992).

The group institutionalized use of its iconic balaclava face masks to avoid identification (Urban, 1992) and the use of gloves to avoid leaving fingerprints and to prevent transfer of weapon and other residues to the volunteer (McGartland, 1997; Urban, 1992). On some operations, members have worn additional external clothing, such as surgical oversocks to cover their footwear, multiple layers of hand protection, including latex gloves, to provide a barrier, and external cloth gloves to provide an absorbent layer to be used, for example, to wipe away sweat on the terrorist's forehead that might otherwise drip and leave telltale DNA evidence.[50] Operatives also reportedly shaved their heads immediately after an operation both to limit the capture of evidence in their hair (Geraghty, 2000) and to ensure a significant change in their appearance in case they were arrested and placed in an identification lineup.[51]

The ways in which PIRA chose to manage operations also contributed to its overall counterforensic effort. For example, although PIRA did file off serial numbers of its weapons to help protect its supply lines (Ryder, 1997), its operational approach to managing its firearms did much more to blunt the effectiveness of ballistics techniques for linking individuals to specific attacks. Because weapons were stored in centralized caches and only delivered to operatives immediately before operations—and taken away immediately afterward—linking a weapon to a specific attack through ballistics did not contribute effectively to prosecution in the same way as linking an individually owned gun to a murder would. This practice represented a marked difference between the utility of evidence of gun ownership or possession against organized terrorist organizations and its value in traditional police investigation of more routine violent crime.[52]

Over time, the group took steps to build practices to destroy or eliminate forensic evidence into its operational plans. These practices

---

[50] Personal interviews with law enforcement officers, Northern Ireland (May 2005).

[51] Personal interviews with law enforcement officers, Northern Ireland (May 2005).

[52] Personal interview with former security forces member, England (May 2005).

included creating the group infrastructure to ensure that members' clothes could be rapidly washed or destroyed—having facilities and people primed to carry out that function—and that the terrorists had a place in which they could quickly bathe. This effort was adapted over time to answer changes in police procedures: In response to the police collecting evidence from washing machines and drainpipes, volunteers were instructed to "always wash clothes by hand and dispose of the water into an outside drain, back garden, or yard" (Geraghty, 2000, pp. 86–87). The group also built organizational capacity to send teams of people to the scene of an operation specifically to "clean" it—destroying as much evidence as possible. For example, after the murder of Robert McCartney in 2005—an unplanned result of an altercation in a pub—an on-call PIRA team went to the scene and washed down the pub with bleach to destroy evidence that might be used against the murderers.[53]

Some of the group's countertactics to forensic science were more direct and violent. In PIRA documents, group leaders suggested that "it may be useful to employ a delaying tactic (such as a hoax bomb or a booby trap) which apart from having obvious military advantages, also allows for time to lapse during which forensic evidence may be dispersed or destroyed" (Geraghty, 2000, p. 84). Incendiary devices and self-destruct mechanisms in bombs were also used specifically to destroy evidence (Bell, 1998; Patrick, 1981). Finally, the group occasionally targeted forensic scientists and their facilities for direct attack: Booby traps were built into "evidence" left at crime scenes ("The Armed Struggle," 1987; Ryder, 1997). And, in an effort to destroy evidence and limit the capabilities available to the security forces, the group directly attacked the Northern Ireland forensic science laboratory—most seriously in 1992, when the group used a multithousand-pound bomb that significantly damaged the facility (Geraghty, 2000).

---

[53] Personal interviews with former law enforcement officer and government security official, Northern Ireland (May 2005).

Conclusion

Throughout its approximately 30-year conflict, PIRA faced a continuously evolving body of defensive technologies fielded against it by British law enforcement and security force organizations. To defend its operational capability and shield itself from these technologies, the group devoted considerable ingenuity and resources to countering the technologies. The group's countertechnology efforts are summarized in Table 5.1.

Table 5.1
Provisional Irish Republican Army Technological Innovations:
Purpose and Intended Mitigation of Government Countermeasures

Innovation	Purpose	Intended Mitigation of Government Countermeasures
Avoiding areas covered by monitoring efforts or technologies	Maintain operational security	Government monitoring of locations
Displacing operations from high- to low-security areas	Avoid risks from security and surveillance	Security and surveillance modes
Avoiding technologies believed to be monitored	Maintain operational security	Government monitoring of communication and other technologies
Obscuring signatures surveillance was designed to pick up through disguise and deception	Maintain operational security	Government monitoring of locations
Separating operatives from weapons	Increase difficulty of holding and prosecuting operatives	Ability to easily tie individuals to violent activities, defeat ballistics approaches tying firearms to specific attacks
Using countersurveillance teams and techniques	Detect and avoid government surveillance efforts	Government surveillance efforts
Using operatives who broke profiles and coercing innocents to carry weapons	Penetrate security cordons	Security around targets or denied areas supported by terrorist profiles or admittance pass systems

Table 5.1—Continued

Innovation	Purpose	Intended Mitigation of Government Countermeasures
Using hoaxes and deception	Directly attack government data-gathering and analysis efforts	Public tip lines and government intelligence databases
Modifying and using alternative explosives detonators	Allow detonation despite electronic countermeasures	Security forces jamming methods to defeat device detonation
Using alternative explosive materials and ingredients	Ensure weapon supply lines	Government control of supplies of explosives
Escalating to larger bombs	Overwhelm defensive hardening measures	Protective mechanisms for high-value or critical targets
Developing shaped charge weapons	Overwhelm defensive hardening measures	Protective mechanisms on vehicle targets
Pursuing alternative (e.g., indirect fire, missiles, sniping) weapons	Attack denied targets	Changes in security forces operations or target protection that made current weapons ineffective
Assembling weapons at targets from small ingredients	Circumvent searches and security intended to keep bombs out of target areas	Security and detection systems
Using bombs and antitamper devices	Directly attack government response capabilities and explosives disposal teams	Bomb disposal capabilities to disrupt operations
Reducing warning times for bombs	Neutralize government response capabilities	Bomb disposal capabilities to disrupt operations
Conducting many simultaneous operations	Saturate government response capabilities	Bomb disposal capabilities to disrupt operations
Staging ambushes on responding teams	Directly attack government response capabilities	Bomb disposal capabilities to disrupt operations, other response capability
Selecting clothing to minimize forensic evidence left at attack scenes	Maintain operational security	Government forensic science analytical capabilities

Table 5.1—Continued

Innovation	Purpose	Intended Mitigation of Government Countermeasures
Laundering or destroying clothing (including specific disposable overclothes) to remove evidence	Maintain operational security	Government forensic science analytical capabilities
Destroying forensic evidence at scenes via secondary devices or "scene clean-up teams"	Maintain operational security	Government forensic science analytical capabilities
Attacking forensic laboratory facility	Directly attack government investigative capabilities	Government forensic science analytical capabilities

From the information available on these activities, three broad conclusions can be drawn that are relevant in examining terrorist countertechnology behavior more generally.

PIRA developed ways to counter a wide variety of technologies. Across all classes of defensive technologies, from surveillance devices to tools for investigating after terrorist attacks, PIRA developed strategies to limit the impact those technologies had on the group. Because we must rely on open-source information, however, it is impossible to give a full accounting of the technologies fielded against PIRA, much less whether PIRA was able to develop counters for each one. Nonetheless, it is clear that the group was able to build sufficient expertise and learning capability to adapt to many different defensive approaches.

PIRA applied varied strategies to develop its countermeasures. In developing its countermeasures, PIRA also showed significant versatility in the approaches it chose to pursue. Even within classes of defensive technologies, the group chose multiple routes, including shifting its operational procedures to neutralize the effects of the technologies, innovating new weapons or technologies or substituting alternative technologies for those it currently used, dodging the technology by displacing its terrorist activities, or attacking the technology directly and seeking to nullify its value or destroy it.

Some of the group's strategies hinged on its ability to gather information about the technologies. Activities such as "challenge-response" experiments, in which the group would test a system to gather information about it or learning about police capabilities and techniques from captured terrorists and their trials, provided information that could be used to shape countermeasures. Many strategies, however, did not rely on such information. For example, PIRA did not need access to internal details of the security forces' vehicle-tracking and computer systems to devise measures that could effectively counter them. In a few cases, PIRA had to devise countermeasures in the absence of the knowledge needed to do so. For example, during a period when the group's bombs stopped detonating as planned, the group had no practical route to determine the source of the problem and had to innovate through trial and error—a process that was costly in time and effort, but one that still proved successful in the end (Collins and McGovern, 1998).

Counterstrategies differed in the burden they placed on the group. Although PIRA developed countermeasures across the full range of technologies that were deployed against it, the development and use of such countermeasures was not without cost[54] to the group. In some cases, the costs were minimal: Ensuring that operatives wore baseball caps to limit the effectiveness of sophisticated CCTV systems made few demands on the group. Other basic operational approaches such as training group members to avoid behaviors that would tip off the authorities were similarly straightforward.

In other cases, addressing the presence of defensive technologies placed significant burdens on the group. Some countermeasures required significant up-front investments in effort and resources—such as PIRA's development of new weapon technologies to counter defensive steps or the group's effort to study coverage of observation posts and surveillance technologies to identify blind spots. In such cases, assuming that the group can make the investments needed to be successful, one action can neutralize the defensive technology going forward, assuming that no changes in its deployment or operation are made in response. However, in some instances, the countermeasures themselves imposed other costs on the group from other perspectives.

---

[54] The term *cost* in this context is intended to be interpreted broadly to include all potential costs, including effort, time, personnel, materiel, and financial costs.

For example, the transition to larger bombs to overwhelm defensive measures resulted in increasing levels of collateral damage around targets, damage that frequently affected groups whose support PIRA was trying to either gain or maintain.

For many of the technologies, however, PIRA's countermeasures required ongoing efforts, adding additional "ingredients" to its day-to-day operations that had to be carried out. Referring to the wide variation of technologies that were aimed at the group, one PIRA member concluded, "Ultimately it is a battle of wits, every operation must be meticulously planned, taking account of the obstacles" (Urban, 1992, p. 118).[55] The need to change vehicles and sites frequently increased the logistical complications faced by the group. Searching buildings, vehicles, and weapons for listening or tracking devices cost the organization time that might have been applied to other pursuits. The need to scout out "clean" automobiles and clone their appearance and identification numbers to evade security forces' vehicle-checking databases required significant effort. When it could be done, it was effective; but it was not always possible to spend the time and effort needed to do it (Urban, 1992). As a result, the net effect of a technology on a group's capabilities depends on its ability to pay the costs involved—for those that cannot, the effect of a defensive technology may be decisive. For others, it may simply provide a "drag" on group performance that reduces its capabilities from where the group would be in the technology's absence.[56]

---

[55] Even with the level of countermeasures that it had available, another PIRA member indicated that the level of technology threat to the group came close to breaking it. Brendan Hughes, a high-ranking member of PIRA, was quoted as saying that the technology "effectively [brought] the IRA to a standstill where it could move very, very little" (see Taylor, 2001).

[56] This is clearly related to the size and pool of resources a group has available. Interviewees suggested that, in PIRA's case, the drag generated by defensive technologies was more important for its operations on the British mainland, which were staged in a more hostile environment and with much smaller, detached operational cells. In contrast, in Northern Ireland, the group had a sufficient pool of human and other resources to accept the costs imposed by the technologies and continue operations (personal interview with law enforcement officers, Northern Ireland, May 2005).

Drawing on PIRA's experience, it is clear that individual technologies are unlikely, on their own, to provide long-lasting answers to the problem of terrorist violence. Such technologies should therefore be viewed as elements of an overall effort against these groups. Even if a deployed technology does not provide a decisive shift in the effort against a terrorist organization, such systems—and the inevitable countertechnology efforts that will occur once they are deployed—can still shape terrorist behavior and limit the capabilities of terrorist organizations.

CHAPTER SIX

Conclusions: Understanding Terrorists' Countertechnology Efforts

The previous chapters examined the countertechnology efforts of four terrorist organizations in four distinct operational contexts. The conflict between the Palestinian terrorist organizations and Israel centered on the use of a comparatively limited number of tactics, with the primary threat to Israel coming from the structures and operatives based in the West Bank and Gaza. JI and its affiliated groups operate across a number of countries in Southeast Asia and, as a result, face a variety of countermeasures fielded by countries whose resources and technological sophistication vary significantly. Similar to Hamas, the components of LTTE's activities in Sri Lanka that were discussed focused on the use of suicide operations staged from a comparatively secure base. Unlike the Palestinian groups, however, LTTE applied that tactic for a markedly different class of terrorist activities aimed at different operational goals and, therefore, generated qualitatively different defensive technology responses. Finally, PIRA represented a case of variety: Because of the length of the conflict in Northern Ireland and variation in PIRA's operations, a profusion of different responses was fielded against the group, requiring that PIRA develop a diversity of countermeasures.

In this chapter, we look across the defensive technologies discussed in these four cases and the range of countertechnology responses developed by the terrorist organizations involved. We describe a set of strategies that terrorist organizations use to defeat defensive technologies, discuss ways in which the transfer of knowledge regarding countertechnology strategies among terrorist groups might affect the utility of defensive technologies, and derive lessons from this analysis that can be used to inform decisions about the acquisition and implementation of defensive technologies in U.S. efforts to improve homeland security.

Terrorist Strategies for Countering Defensive Technologies

The actions taken by terrorist organizations in response to governmental deployment of defensive technologies vary from basic displacement of terrorist activities—for example, shifting to softer targets when others are protected—to sophisticated technology development efforts to, for example, evade or deceive surveillance systems. This variety in countertechnology actions is, no doubt, a function of the wide range of defensive technologies that governmental authorities have used to try to prevent acts of terrorism and the differences in the goals, activities, and operational environments of terrorist organizations.

Despite this variety, however, the many specific countermeasures terrorist organizations adopted have common elements that permit us to reduce this behavioral diversity to a smaller number of fundamental countertechnology strategies. In considering the response of terrorist organizations to the deployment of these technologies, their counter-efforts can be broken down into four main classes:

1. **Altering operational practices.** By changing the ways it carries out its activities or designs its operations, a terrorist group may blunt or eliminate the value of a defensive technology. Such changes frequently include efforts to hide from or otherwise undermine the effect of the technologies.

2. **Making technological changes or substitutions.** By modifying its own technologies (e.g., weapons, communications, surveillance), acquiring new ones, or substituting new technologies for those currently in use, a terrorist group may gain the capacity to limit the impact of a technology on its activities.

3. **Avoiding the defensive technology.** Rather than modifying how it acts to blunt the value of a defensive technology, a terrorist group may simply move its operations to an entirely differ-

ent area to avoid it. Such displacement changes the distribution of terrorism and, although this may constitute successful protection in the area where the defensive technology is deployed, the ability to shift operations elsewhere limits the influence the technology can have on the overall terrorist threat level.[1]

4. **Attacking the defensive technology.** If appropriate avenues are available, a terrorist group may seek to destroy or damage a defensive technology to remove it as a threat.

This taxonomy of fundamental strategies adopted by terrorist groups provides a structured way to assess the countermeasures that a group might use when challenged by a new defensive technology. However, although it is useful to break down the full range of terrorist counterstrategies into a limited number of classes, doing so should not suggest that, when faced with a novel security challenge, a terrorist organization will necessarily limit itself to selecting one of the four types of countermeasures. In the history of the groups described here, there is a number of instances in which multiple strategies were fielded either simultaneously or consecutively against one defensive technology. For example, to counter border controls in the countries in which they operated, JI and its affiliate groups used false documents to deceive the systems (strategy 1), used different transportation modes in which the measures were less stringent (strategy 3), and attempted clandestine crossings where they would not face the technologies at all (strategy 3).

Furthermore, for some defensive technologies, unambiguously placing a group's countereffort in one of these four classes may be difficult, since that effort may involve a combination of strategies. For example, to counter surveillance equipment, PIRA fielded its own detection technologies (strategy 2) that allowed it to locate and remove the devices (strategy 4). The efforts of Hamas and PIRA to develop or acquire mortar technologies so they could attack their targets from

---

[1] Enders and Sandler (2004) have done comprehensive analyses of such behavior for transnational terrorism, including displacement effects among target classes as a result of deployment of defensive technologies.

above—thereby avoiding defensive hardening—combines technological change (strategy 2) and avoidance (strategy 3).

Although such cases could be viewed as the consecutive application of two different strategies to the same defensive technology, such combination efforts emphasize the difficulty of cleanly breaking real-world behaviors into distinct orthogonal categories. Reflecting this complexity, the four strategies are likely better viewed as the extremes of four overlapping sets (Figure 6.1) in which specific counterefforts may draw on one or more of the basic strategies in an effort to defeat or circumvent a specific defensive technology.

The strategies adopted by the organizations discussed in this book differed across the classes of defensive technologies. Here, we summarize our observations regarding the responses of terrorist organizations to each of the five major defensive technological approaches discussed in the introduction.[2]

Information Acquisition and Management

In general, terrorist groups changed their operational practices in response to surveillance or sought to avoid the technologies completely. For example, they frequently used different types of camouflage or deception to hide from such technologies or to obscure the behavioral or other signature the technologies were primed to detect. Palestinian groups in the West Bank and Gaza applied such techniques in an effort to avoid Israel's attempts to target their members and preempt their offensive activities. The groups we studied also used compound strategies, combining, for instance, modifying or substituting technologies for those they were using to allow the group to avoid surveillance

---

[2]   As stated in the introduction to this book, limits on the availability of information on the deployment of defensive technologies and terrorist responses, whether for security or other reasons, could skew the results of analyses such as this. Although the research team has pursued a variety of routes to gather data on the terrorist groups examined in the study and the defensive technologies used against them, there are almost certainly technologies and counterefforts that are not reflected here. As a result, the following summaries provide overviews of the trends in the counterstrategies adopted by the groups, illustrated with examples drawn from the earlier chapters. As the nature and scope of any "missing data"—technologies and counterefforts of which the team is unaware—is unknowable, we have not attempted to quantify the data or to provide more specific breakdowns.

Figure 6.1
Terrorist Countertechnology Strategies

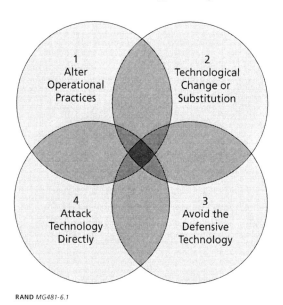

systems. Examples of this behavior included efforts to mask explosives with other materials or coatings to fool detection devices and the
substitution of novel communication modes for modes thought to be
easily monitored. There were many fewer examples of groups attacking surveillance systems directly. However, when it was possible (e.g.,
PIRA's use of tip line systems to bait security forces into traps or impli-
cate innocent citizens as group members), the payoff to the group was
significant.

Preventive Action
Taken together, the groups we studied used all four available strategies to counter efforts to degrade their capabilities. Hamas changed
its operational practices by increasing secrecy to protect its assets from
Israeli targeting; in other groups, changes in attack planning to avoid
jamming of cell phone detonators provided ways to counter some technologies. In response to efforts to curtail its money flows, JI sought to
go through the Philippines and Thailand, which were more permissive,

or to use informal *hawala* or other transfer systems to avoid the risk its finances would be compromised. In response to security efforts to jam its detonation modes and constrain the availability of weapons and explosives, PIRA made technical changes and substitutions—modifying its detonators to defeat the technologies and manufacturing its own weapons when needed. Although yet unused, Hamas reportedly obtained antiaircraft systems to attack directly the air platforms Israel has used in its targeting of key Hamas members and assets.

Denial

The most potent strategies for countering technologies that harden targets against attack—and, in some ways, the strategy that placed the least burden on the terrorist—were operational changes that allowed penetration of target defenses. When it became clear that security and police forces were using terrorist profiling to detect operatives, every group sought and used terrorists with characteristics that were inconsistent with the profile and could therefore avoid detection. PIRA, in its use of so-called proxy bombers—innocents compelled through threat to carry a bomb to its target—took this concept to its extreme. Groups also substituted new technologies or modified their existing technologies, choosing alternative attack modes or scaling up their weapons to defeat or overwhelm defenses. Lastly, groups sought to avoid defenses entirely by choosing different targets in different areas or shifting to new target sets (e.g., PIRA shifting from bombings to sniper tactics).

Response

Of the terrorist groups examined for this study, PIRA was the primary organization that made a conscious effort to counter government's ability to respond and mitigate the effects of attack operations. Its approach relied on changes in operational practices coupled with technical changes—for example, modifying the way bombs were constructed and used to limit security forces' ability to defuse them before detonation. This area also includes an example of what might be labeled an "n+1" attack to neutralize response measures: Having determined the number of response teams the government had available ("n"), the group staged a combination of bombs and hoaxes that were certain to exceed that capability. The group also directly attacked response teams through the use of secondary devices placed at the scene of attacks.

Investigation

The preceding chapters contain a fairly limited number of examples of counterefforts aimed at investigative technologies. Groups that did seek to counter these technologies focused on changes in operational procedures—before, during, and after operations—to limit the likelihood that forensic investigators could identify the terrorists involved. In a few instances, terrorists attacked these capabilities more directly with bombs aimed at injuring investigators and destroying evidence or facilities.[3]

Transferability of Terrorist Countertechnology Strategies

To generate countermeasures for defensive technologies, terrorist groups usually must expend time and effort. That development process—during which the organization must assess the defensive technologies and shape its response—provides a window of time when the technologies are effective, during which even technologies that are eventually neutralized produce some payoff. The duration of this window of effectiveness will depend on the nature of the technology—what is required to counter it—and on the nature of the terrorist group and its available capabilities. If a group must develop new countermeasures from scratch, how rapidly it can do so will depend on how well the group learns and implements the results of its learning efforts (see Jackson et al., 2005a, 2005b). To the extent that counterstrategies that have already been proven successful can be transferred—for example, from one area of a group's theater of operations to another or among different terrorist organizations—the need for lengthy learning efforts can be eliminated, and that window of effectiveness can significantly

---

[3] Planting explosives in fake evidence at crime scenes—with the intent of detonating them inside police or forensic facilities—is analogous to using secondary devices targeting responders at the incident scenes.

shrink or even close entirely. Understanding the factors that affect the transferability of such strategies is therefore important in planning homeland security operations and in the design of defensive technologies to support those efforts.

The determining factor in the threat posed by specific counter-technology approaches of terrorists is whether they provide generally useful ways to undermine the effectiveness of such systems or are specific to the context in which they were originally developed. To the extent that strategies are context-specific, their transferability and potential utility for other organizations or in other theaters may be quite limited. The clearest example of a context-specific counter-technology action in our study is the intentional use of satellite telephone communication modes by LTTE, which was prompted by its knowledge that the Sri Lankan government lacked the capability to monitor such systems.[4] Across several of the groups, use of particular concealment approaches was similarly context-specific; though the general concept of concealing activities from detection systems and surveillance approaches is certainly transferable, the tactics needed to do so depend on the characteristics of the systems deployed in each area.[5] Constraints in a group's environment can also produce context-specificity for otherwise general countertechnology strategies. For example, PIRA increased the size of its weapons to overcome target hardening, which resulted in significant collateral damage around the group's intended targets. Because much of that damage hit populations whose sympathies the group sought to maintain, this counterstrategy was less than ideal for the group's operations.[6] LTTE faced this same

---

[4] This contrasts, for example, with the U.S. capability to monitor such phones, which was broadly reported in stories about U.S. monitoring of Osama bin Laden's satellite telephone.

[5] For example, JI worked through local terrorist organizations rather than performing certain activities for itself (e.g., in the Philippines) because those groups were better integrated into their local context. The availability of this strategy depended on the availability and willingness of such groups and, therefore, would not necessarily always be possible.

[6] Personal interviews with law enforcement technical experts, Northern Ireland (May 2005).

problem and, as a result, did not pursue this strategy because it did not want the increase in collateral damage that would result.[7]

To the extent that approaches are not specific to the context in which they were developed, the likelihood that they can and will be transferred among different terrorist organizations depends on characteristics of the countertechnology approaches.[8] The specific issue of such transfer among terrorist groups is the topic of a companion study performed concurrently with this project. The report of that study, *Sharing the Dragon's Teeth: Terrorist Groups and the Exchange of New Technologies*, addresses in detail the full range of factors affecting the potential for successful technology and knowledge transfer, including the characteristics of the source and receiving organizations, the transfer mode, and the technology or knowledge itself. To assess the potential for successful transfer of specific counterstrategies between specific terrorist organizations, the full range of characteristics described in the companion document would have to be considered. Because the present analysis addresses the counterstrategies, the following discussion

---

[7] This analysis examined only "one round" of measure-countermeasure interaction between terrorist groups and organizations seeking to combat terrorism. In multiround interactions, there would likely be countermeasures fielded to the terrorists' countertechnology strategies, then counter-countermeasures, and so on. At some point, such a multiround interaction could reach a point at which one side or the other exhausts all available or practical routes for adaptation. In others, such an endpoint may not exist.

Such multiround interactions could make otherwise highly transferable countertechnology strategies far less useful for other terrorist groups. For example, the use of false identity documents to circumvent systems that rely on such documents to substantiate individuals' identities was used by many groups and is clearly readily transferable. The use of systems to detect such forged documents in some areas would reduce the utility of the strategy—essentially forcing context-dependence onto it for areas that do not use such secondary systems.

[8] Although not the primary focus of this effort, an interviewee cited an example in which the legal framework surrounding security efforts can create context-dependence in terrorists' counterstrategies. In Northern Ireland, UK law permitted holding suspects for seven days without charges, during which they could be interrogated. PIRA's counterinterrogation approaches focused entirely on teaching its members techniques for resisting questioning for that seven-day period. In another country where detentions could be extended, their counterstrategies would be much less applicable (personal interview with government security official, Northern Ireland, May 2005).

will focus on their characteristics and their effect on transferability of the strategies among terrorist organizations.

In many cases, the countertechnology strategies used by the groups we studied were comparatively simple innovations and, therefore, relatively straightforward to transfer. It is significant, for example, that counterstrategies focused on avoiding specific communication modes, maintaining certain basic operational security practices, and using terrorist operatives who did not fit the profiles used by security forces to identify targets were observed in all four of the groups examined for this study. Instructions to carry out such countertechnology strategies could be readily captured in written form and transferred among individuals or from group to group either physically or via electronic modes such as the Internet.[9] Other approaches that can be captured in explicit form[10]—particularly those driven by groups adjusting their operational practices to neutralize a technology—could also be readily transferred. Counterforensic activities, for example, although not observed across all of the groups examined here, have been described in instruction manuals developed by PIRA; the Earth Liberation Front, an environmental terrorist organization, has captured such lessons as part of training materials that have been broadly disseminated on the Internet (Trujillo, 2005; Jackson et al., 2005a). In such cases, the barriers to spreading technology are quite low.

In other cases, implementing counterstrategies was more complex. Strategies that require the acquisition or development of specific technologies depend on groups having the ingenuity to build them or the right contacts and resources to obtain them. For example, obtaining new weapon systems and learning to use them effectively—as PIRA and Hamas did—is frequently more difficult than making basic operational changes. Similarly, modifying technologies that are currently in

---

[9] For example, similar topics and countermeasures are covered in captured manuals of jihad produced by al Qaeda and shared over the Internet.

[10] See Jackson (2001) and Jackson et al. (2005a) for a more complete discussion of the differences between *explicit knowledge* (knowledge and technology that has been captured or embodied in physical form) and *tacit knowledge* (knowledge that is difficult or impossible to capture in that way) and the implications of the difference for its transfer among individuals or organizations.

use to evade countermeasures or attempt to reconstitute performance requires a level of technical expertise, resources, and the ability to experiment to ensure that the modifications match the local operational environment. Transfer in these more complex cases is subject to all of the issues associated with technology and knowledge transfer that can greatly reduce how effectively they can be moved from one organization to another.[11]

Implications of Terrorist Countertechnology Activities for Homeland Security Efforts

From the perspective of homeland security technology planning, the historical record of terrorists' efforts to counter defensive technologies is not encouraging. Although there are likely technologies that the groups examined in this study have been unable to circumvent— about which information is not available in the open literature—they were able to develop counterstrategies for a wide variety of technologies, demonstrating significant ingenuity and adaptability. As a result, for most technologies, the groups will adapt to circumvent them, and security organizations will have to respond, starting the measure-countermeasure cycle again.

This observation is echoed in comments from a number of interviewees, who said that technology in and of itself cannot provide a solution to terrorism—no technological "silver bullet" exists—but that it provides only an additional advantage in concert with good human intelligence and investigative efforts. The advantage provided by technologies is at its greatest before adversary groups have had the chance

---

[11] For example, although PIRA demonstrated considerable ability to acquire new weapons to circumvent hardening of potential targets, even the transfer of that group's technologies to current "dissident republican groups" (including the Real IRA and the Continuity IRA), which drew parts of their membership from PIRA, has not been totally straightforward. Interviewees indicated, for example, that some such groups have been sliding back to "lower-tech" modes of operation, since they have not been able to carry out higher-technology operations successfully (personal interviews with law enforcement technical experts, Northern Ireland, May 2005).

to develop counterstrategies—and, depending on the groups involved, that advantage can be fleeting.

Beyond providing an important caution and perspective about what security organizations and the public should expect from defensive technologies, the experience of these organizations also provides relevant lessons to inform homeland security planning and improve future generations of defensive technologies. These lessons address ways to enhance the functioning and robustness of future defensive technologies and approaches to improve planning for the technological components of homeland security efforts.

Lessons for the Design of Defensive Technologies

To ensure that new defensive technology systems provide the greatest potential security benefits, they must be designed with terrorist countertechnology behaviors and past successes in mind. The efforts of the groups studied here suggest three techniques or approaches to use in developing plans for new defensive technologies.

Red teaming technology systems. Given terrorist countertechnology behaviors, there is a clear need to test or "red team" new technologies to draw on the terrorists' available palette of counterstrategies (e.g., Figure 6.1) to assess a technology's limits before it is built and deployed. Such testing is established practice for many security technologies and practices. Experience with the groups described here, however, suggests that such testing is needed even for technologies that adversaries "do not see"—such as information collection, processing, and database technologies to which an adversary may never have direct access. This testing can help to identify potential routes that terrorists might use to circumvent or degrade their functioning.

Assessing adversary information requirements. In the design of new defensive technologies, there is an obvious value in analyzing the information an adversary would need to circumvent the defensive technology and how the adversary might gain access to that information. Security organizations should consider, for instance, whether the efficacy of the technology hangs on the ability to "keep secrets" about how it functions, how such secrets might be compromised, and whether the group could discern them from the outside.[12] Analysts should also consider whether testing—such as action-reaction challenging of the system by adversary probes—could provide the needed data, whether groups that are willing to sacrifice low-level operatives in exploratory operations against the system can learn what they need to know, and whether the characteristics of the system are transparent enough that it is clear how its capabilities might be saturated (i.e., how to craft a successful "n+1" attack to overcome the technology).[13] To the extent that features can be built into the system that defeat or degrade the ability to gather the needed information, the ability of the technology to deter or defeat terrorist operations will be bolstered.

Designing flexibility into defensive technologies. The observations presented suggest that, for most defensive technologies, terrorist organizations will eventually develop counterstrategies that limit their value. As a result, systems that are flexible—that are not locked into specific modes of operation—provide an added value. If the characteristics of a system are essentially fixed, it is a static target for terrorist adaptive efforts and, once compromised, may provide little security benefit. This notion suggests that focusing on ways to build flexibility into defensive systems could be valuable. For example, just as changes in operational practices provide terrorists with a variety of ways to get around technologies—obscuring the signatures they were designed to detect, using deception, adjusting the speed or character of their operations—they could similarly provide a variety of strategies for altering the character of defensive systems. Changes in hardening procedures or guard activities are common examples of the integration of operational shifts to disrupt terrorist efforts to circumvent defenses. Taken further, changes that are activated when the early stages of terrorist

---

[12] Techniques such as deception could be used to help protect relevant characteristics of countermeasure systems as well.

[13] How an adversary might build up the needed information requires foundational knowledge of how its organizational units operate and the cultural influences that may shape how its members pursue their goals. The difficulties in building such understandings for groups drawn from very different cultural traditions are broadly appreciated.

activities are detected—strategies to "dynamically harden" facilities of concern[14]—could contribute significantly to efforts to derail terrorist efforts to circumvent security measures.

Anticipating how technologies will guide terrorist adaptation. Deployment of new defensive technologies is one of a variety of elements that can force a terrorist conflict from a static to a dynamic frame, with each side forced to change and respond to the actions taken by the other. Being able to anticipate how an adversary will likely respond to a specific shift in tactics or strategy is critical for shaping effective countermeasures and achieving success in such conflicts. Terrorists have long recognized the importance of this ability to anticipate and shape behavior (e.g., Bell, 1976); it is no less important for security and homeland security planning.

These efforts to anticipate terrorists' moves and countermoves against defensive technologies are particularly important because a group's efforts to adapt and survive when faced with a new technology can help to build it into a more potent threat than existed before the technology was deployed. The most basic manifestation of this effect is the selective pressure that technologies and other security measures exert on terrorist groups, eliminating the less talented or professional individuals and reducing a group to a hardened core.[15] Beyond such unavoidable effects, technologies can shape terrorist capabilities in more damaging ways. For example, if a particular security measure pushes terrorists toward weapons and attack modes for which no good defensive options exist, the end state threat level may be higher than before the measures were introduced. Similarly, how terrorist organizations respond may change the nature of the threat they pose. For example, the challenges that Hamas and the other Palestinian groups

---

[14] The shifts in security posture in specific areas or patrol routes used by the security forces in Northern Ireland to disrupt PIRA operations are a basic example of such dynamic activity. At specific facilities, dynamic strategies could include activation of varied sets of traffic barriers within an otherwise static perimeter and partial shutdown of components of the facility—which would significantly change the target's nature of the target, its security, and profile—and could therefore pose significant disruption to terrorist efforts to learn and circumvent facility defenses.

[15] Personal interview with former security forces member, England (March 2004).

faced from Israeli security measures provided an incentive for them to strengthen their ties with Hizballah as a source of expertise and technology to solve their problems; the potential effects of those strengthened ties may go well beyond their countertechnology activities.

To limit the potential for unintended consequences in terrorist adaptation, the design process for defensive systems should explore the effect of terrorist countertechnology responses not only on the value of the defensive system, but also on the group overall and the nature of threat it poses. Although the potential for these groups strengthening as a result of their defensive activities is not an argument for inaction, the actions taken by security organizations should be designed to minimize the chances for such consequences as the dynamic conflict with such groups continues.

Lessons for Planning the Technological Components of Homeland Security Efforts

When terrorists are successful in countering all or part of the functioning of a defensive technology, the utility of the system may be significantly reduced or lost entirely. Such losses devalue the costs society pays to design, produce, field, use, and maintain the technology[16]—where the concept of costs includes not only financial and materiel costs but also auxiliary costs such as any reductions in privacy and civil liberties or costs paid in time or inconvenience by the public as a result of implementation of the security measures.[17] The efforts of these terror-

---

[16] For a nation as large and populous as the United States, these costs can be considerable. For example, at the time of this writing, major initiatives regarding border security and critical infrastructure protection are under consideration. Given the scope of both problems and the resources needed to implement solutions, considering how terrorists might act to counter protective measures that are put in place is clearly critical.

[17] The costs—of all types—that society is willing to pay for a new defensive technology are obviously context-dependent. Costs that might have seemed perfectly reasonable during the height of PIRA activities in Northern Ireland for even transitory advantage might be wholly unacceptable in other nations and other contexts. The following discussion does not address the levels of these costs in particular; instead, it focuses on the need to include the potentially transitory nature of the advantage provided by defensive technologies in the cost-benefit assessment of the technologies.

ist organizations suggest three key lessons to be taken into account in planning and implementing security measures.

Include terrorist countertechnology efforts in programmatic and cost-benefit analyses of defensive systems. In assessing a novel or enhanced technology and the costs associated with producing and fielding it, the risk that it will fail to deliver within its planned budget is an established component of management planning. Like the competitive risk that another firm will develop a superior product, rendering a company's investments meaningless when both reach the market, successful terrorist countertechnology efforts can similarly destroy the competitive advantage of a new or enhanced defensive system. This countertechnology risk must be assessed and included as part of program management above and beyond the technological and other risks inherent in the effort.[18]

The level of additional risk that must be managed will largely be determined by the nature of the technology and the counterstrategies available to the adversary. Security experts will need to determine whether simple operational changes will suffice to counter a new technology or whether more complex measures are required. In addition, analysts must consider whether a single counterstrategy will eliminate all benefits of the technology or whether it has elements that must be countered separately. Even if the terrorist develops counterstrategies, it is important to consider whether implementing them requires only a one-time cost or whether it requires the terrorist group to commit extra effort every time it takes action and therefore will maintain a drag on its capabilities and resources. Finally, security planners must determine whether the technology is frozen into a single configuration or whether it is flexible, so it can be changed when terrorists threaten to circumvent it. In each case, the latter options involve less risk. As a result, their benefits would have to be discounted less when assessing the net costs and benefits of a potential defensive system.

Consider the relative costs of countering a technology and the cost of the technology itself. The cost that a defensive technology can

---

[18] See Chow et al. (2005) for an example of an analysis of a defensive technology that explicitly considers countertechnology risks

impose on a terrorist group—in effort and resources required to either withstand or counter its effects—is one measure of its value. If the cost is great enough, the technology's effect can be decisive. The cost that the nation should be willing to pay for a technology system must be related to its potential effect on its adversaries. When a technology can be countered with little investment on the terrorist's part, the balance is in the terrorists' favor. In such a case, the price that society should pay for the technology should be very low. Situations in which expensive technologies can be countered by low-cost countermeasures are particularly adverse.[19]

Address "multistep" countertechnology activities in assembling groups of security technology investments. Although this discussion focuses predominantly on single-step interactions between terrorist groups and defensive technologies—a single response by a group to a deployed technology—real conflicts are multistep contests. In consecutive iterations of measure and countermeasure competition, the potential exists for terrorists to eventually overwhelm even the most adaptable defensive technology and reduce it to uselessness. If and when that occurs, new options will be needed. Given the potential for such "adaptive destruction" of individual security approaches, planning must consider a variety of defensive technology options, maintaining possibilities for alternative approaches in the event that currently effective technologies are neutralized. If decisions are made to pursue a specific path, the costs of maintaining other technologies in reserve—perhaps not fully developed, but at a stage at which they might be called on if needed—should be considered as well. Such an approach is analogous to maintaining a diversified portfolio of investments, containing a variety of options, where comparatively small investments provide various hedges against different shifts in circumstances.

---

[19] Interviewees focused on the example of baseball caps being used to counter CCTV cameras in this context. Not only did an inexpensive countermeasure defeat the camera system, but also any additional investment in capability (e.g., facial recognition) that was added to the system (personal interviews with former law enforcement member, England, May 2005). This conclusion assumes that modifications in the technology (the flexibility in application and usage discussed above) are not possible to address the countermeasure.

In Conclusion:
The Role of Technology in Combating Terrorism

Although technologies can provide an edge in the effort to combat terrorism, that edge can be dulled by terrorist countertechnology efforts. The potential of these efforts to degrade the effectiveness of defensive systems means that they must be addressed in planning to ensure that efforts to protect society are effective. Particularly now, when a stated goal of some terrorist groups is to inflict economic damages on the nations they target, ensuring that resources are used effectively takes on even more importance. Expending resources for systems that can be easily neutralized in a sense "does the terrorists' work for them" by diverting those resources away from better uses inside or outside the security arena.

Beyond identifying a number of elements that should be considered in decisionmaking regarding defensive technologies, however, the history of the four terrorist groups examined in this book also provides a more complete and nuanced view of technology's role in combating terrorism. Although the performance of most technologies will eventually be degraded at least in part by countertechnology activities, a broader view of technology's role in homeland security efforts provides ways to bolster their impact even in the face of inevitable challenges.

In assessing the effects of defensive technologies, "defense" need not be viewed as an either-or proposition. There are technologies that, once a countermeasure is developed, can essentially be ignored by terrorists. However, others can continue to pose problems for these organizations even after they know how to evade or neutralize them. Those problems are a price the group must continue to pay over time—in the effort needed to counter the technology, the increased planning burden it creates, new or different weapons that must be procured, or resources that must be expended to protect the group from its effects. The lasting presence of even a countered technology can also increase a group's operational risks; for example, even if the terrorist group knows how to evade forensic investigation technologies, it must execute those counterstrategies effectively on each and every operation, or the group will immediately face the full effect of those technical capabilities.[20] In this view, the value of a defensive technology is not necessarily that it can exact a high enough "one-time price" on a terrorist organization to overwhelm it, but that the technology is a drag on the terrorist group's operations over time and the cumulative costs gradually wear away the group's capabilities and operational freedom.[21] This reality should shape how success is measured in security efforts and programs.

Similarly, even if the terrorist group develops ways to counter a technology, the deployment of defensive systems provides a route to shape the behavior of terrorist organizations over the longer term. Such an approach looks at effects of technologies not from a limited "defensive impact" perspective, but from a more holistic view of the total influence they can exert. Countertechnology efforts are, in fact, one way in which such systems shape terrorist behavior—for example, to the extent that they divert groups' "effort budget" away from offensive and toward defensive activities, that in and of itself can be valuable. Such shaping can take other forms. Defensive technologies might be used to divert terrorists away from particular weapons and toward others; toward operational behaviors that are more systematic and, therefore, more easily monitored; or into activities that have more obvious signatures for detection and disruption. To the extent that such effects can be foreseen in defensive technology design, such shaping can provide durable benefits even if individual terrorist groups learn how to defeat the technologies themselves.

---

[20] Depending on the specific characteristics of the technology, one can envision systems that "function even as they fail." The ways in which groups neutralize them and how those methods are spread within the group provide information and insight into the group's capabilities and activities even as it is in the process of defeating the technology. In considering protective strategies in which security or other measures are adjusted based on observed terrorist behaviors, such approaches could provide valuable data to underlie adaptive defense efforts.

[21] Whether a group can simply pay the price exerted by a technology will depend on the resources it has available—in people, materiel, finances, and so on. For example, members of security organizations interviewed for this research indicated that the "drag" imposed on PIRA by defensive technologies was much more serious for its England campaign (that relied on small cells of individuals supported by comparatively limited infrastructure) than for its operations in Northern Ireland, where the group had more personnel and other resources (personal interview with law enforcement members, Northern Ireland, May 2005).

An understanding of past terrorist efforts to counter defensive technologies underscores the complexity of designing new systems to protect society from the threat of these violent organizations. This analysis suggests that, in designing protective measures, it should not immediately be assumed that the newest and most advanced defensive technologies—the highest wall, the most sensitive surveillance—will best protect society from terrorist attack. Drawing on common metaphors for defensive efforts, a fortress—relying on formidable but static defensive measures—is a limiting strategy. Once a wall is breached, the nation is open to attack. Depending on the adaptive capabilities of the adversary, a defensive model built of a variety of security measures that can be adjusted and redeployed as their vulnerable points are discovered provides a superior approach to addressing this portion of terrorist behavior. However, whatever combination of models and measures is chosen, it is only through fully exploring our adversaries' countertechnology behaviors that vulnerabilities in our defenses can be discovered and the best choices made to protect the nation from the threat of terrorism.

Prominent Acts of LTTE Suicide Terrorism, 1987–2002

Table A.1

Prominent Acts of LTTE Suicide Terrorism, 1987–2002

Date	Target	Purpose	Remarks
1987	Tamil University taken over by SLAF	Destroy strategic military location	Attack modeled on the 1983 Hizballah truck bombing in Beirut; 75 people died in the assault.[a]
1991	Rajiv Gandhi (Indian Prime Minister)	Assassinate VIP	Gandhi was assassinated for his decision to curtail Indian support for LTTE and lead a peacekeeping force to stabilize the situation in Jaffna. This is the only act of concerted terrorism that LTTE has carried out beyond the Sri Lankan theater. Eleven others were killed in the attack.[b]
1991	Joint Operations Center (JOC), Ministry of Defense	Destroy strategic military location	The blast killed more than 20, wounded 50, and destroyed vehicles as far away as 300 yards from the JOC premises.[c]
1993	Ranasinghe Premadasa (Sri Lankan President)	Assassinate VIP	Premadasa was killed by a deep penetration mole who had been on the presidential staff for several years. He was targeted for his endorsement of the 1987 Indo–Sri Lankan Peace Accord. The attack killed 17 and wounded more than 60.[d]

Table A.1—Continued

Date	Target	Purpose	Remarks
1994	Gamini Dissanyake (opposition leader contesting the 1994 presidential elections)	Assassinate VIP	Dissanyake was targeted for his key role in arranging the details of the 1987 Indo–Sri Lankan Accord; an additional 50 people were killed in the attack (which bore strong resemblances to the Gandhi assassination).[e]
1995	Naval gunboats (SLNS *Suraya* and SLNS *Ranasuru*)	Destroy strategic naval asset	Both ships were completely destroyed in the twin assaults, which left 11 sailors dead (the two ships were berthed with skeleton crews at the time of the strikes). It has been speculated that al Qaeda's attack on the USS *Cole* was modeled on this operation.[f]
1995	Ceylon Petroleum Corporation oil facility	Destroy strategic economic target	Four oil storage tanks were destroyed, triggering one of the largest fires ever in Colombo. Twenty-one persons were killed in the operation.[g]
1996	Central Bank	Destroy strategic economic target	This is the most destructive act of terrorism to have ever taken place in Sri Lanka, killing 91 and injuring more than 1,400.[h]
1997	Colombo World Trade Center (WTC)	Destroy strategic economic target	The WTC was hit just one week following its inaugural opening. The attack, which killed 15 and injured more than 100, was thought to be in retaliation for the U.S. decision to designate LTTE as a terrorist organization (the bombing is one of the few conducted by the Tigers that has made no attempt to limit foreign casualties).[i]

Table A.1—Continued

Date	Target	Purpose	Remarks
1999	Chandrika Kumaratunga (Sri Lankan president)	Assassinate VIP	Kumaratunga was targeted for her hard-line stance against LTTE and (then) refusal to negotiate with the group. Although the president survived the attack, which was carried out by a male BT dressed as a woman, she suffered damage to her face and lost her right eye. Fourteen other people were killed, including a top officer in charge of Kumaratunga's security.[j]
2001	Bandaranaike International Airport	Destroy strategic economic target and hub of critical transportation infrastructure	Twenty-six civil and military aircraft were destroyed in the attack; it is estimated that losses to Sri Lankan Airways exceeded $350 million.[k]
2001	Oil tanker	Destroy economic target	Attack involved a coordinated strike force consisting of five suicide boats[l]

[a] Jayasinghe (1996), Waldman (2003).

[b] "Tigers Suspect in Gandhi Assassination" (1991), "Tiger Terror" (1995).

[c] de Silva (1991).

[d] Jayasinghe and Ahamath (1993), "Tiger Terror" (1995).

[e] "Gamini Killed in Bomb Blast" (1994), "Tiger Terror" (1995).

[f] Western diplomatic official, Sri Lanka (May 2004); Senanaysake and Candappa (1995); Senanayake (1995).

[g] Malalasekera (1995).

[h] Yapa (1996), Jayasinghe (1996).

[i] Ellatamby (1997), Stackhouse (1997), Burns (1997).

[j] "Wounded Sri Lankan President Calls on Tamils to Join Fight Against Terrorism" (1999), Jayamaha (1999).

[k] Gunaratna (2001), "The Tigers Pounce" (2001), "Tamil Rebels Raid Sri Lankan Airport" (2001).

[l] Luft and Korian (2004).

Bibliography

Abraham, Thomas, "The Emergence of the LTTE and the Indo-Sri Lankan Peace Agreement of 1987," in Kumar Rupesinghe, ed., *Negotiating Peace in Sri Lanka: Efforts, Failures and Lessons*, London: International Alert, 1998.

Abu'Amr, Ziyad, *Islamic Fundamentalism in the West Bank and Gaza: Muslim Brotherhood and Islamic Jihad*, Bloomington: Indiana University Press, 1994.

Abuza, Zachary, "Funding Terrorism in Southeast Asia: The Financial Network of Al Qaeda and Jemaah Islamiya," *Contemporary Southeast Asia*, 2003a, Vol. 25, pp. 169–199.

———, *Militant Islam in Southeast Asia: Crucible of Terror*, Boulder, Colo.: Lynne Rienner, 2003b.

———, "Al Qaeda in Southeast Asia: Exploring the Linkages," in Kumar Ramakrishna and See Seng Tan, eds., *After Bali: The Threat of Terrorism in Southeast Asia*, Singapore: Institute of Defense and Strategic Studies, Nanyang Technological University, 2004, pp. 133–160.

Adams, James, Robin Morgan, and Anthony Bambridge, *Ambush: The War Between the SAS and the IRA*, London: Pan Books, 1988.

Almog, Doron, "Lessons of the Gaza Security Fence for the West Bank," *Jerusalem Issue Brief*, Vol. 4, No. 12, December 23, 2004. As of July 1, 2005: http://www.jcpa.org/brief/brief004-12.htm

Almonte, Jose T., "Enhancing State Capacity and Legitimacy in the Counter-Terror War," in Kumar Ramakrishna and See Seng Tan, eds., *After Bali: The Threat of Terrorism in Southeast Asia*, Singapore: Institute of Defense and Strategic Studies, Nanyang Technological University, 2004, pp. 221–240.

Anderson, John, and Molly Moore, "Hamas Leader Killed in Gaza; Founder of Militant Palestinian Group Is Targeted in Israeli Airstrike," *The Washington Post*, March 22, 2004, p. A1.

"The Armed Struggle," *Iris*, Vol. 1, No. 11, October 1987, pp. 28–41.

"Azahari's Blinking Cell Phone," *Tempo*, November 17, 2003, p. 15.

Baker, John C., "Jemaah Islamiyah," in Brian A. Jackson, John C. Baker, Peter Chalk, Kim Cragin, John V. Parachini, and Horacio R. Trujillo, *Aptitude for Destruction*, Vol. 2, *Case Studies of Organizational Learning in Five Terrorist Groups*, Santa Monica, Calif.: RAND Corporation, MG-332-NIJ, 2005, pp. 57–92. As of December 27, 2006:
http://www.rand.org/pubs/monographs/MG332/index.html

Barel, Zvi, "Newspaper Essay by Israeli Political Commentator, Discussing the Change in Tactics of Palestinian Militants in Gaza," *Federal News Service* (Ha'aretz), May 13, 2004.

Barker, Alan, *Shadows: Inside Northern Ireland's Special Branch*, Edinburgh: Mainstream Publishing, 2004.

Barzilay, David, *The British Army in Ulster*, Vol. 1, Belfast, Northern Ireland: Century Books, 1973.

———, *The British Army in Ulster*, Vol. 2, Belfast, Northern Ireland: Century Books, 1975.

———, *The British Army in Ulster*, Vol. 3, Belfast, Northern Ireland: Century Books, 1978.

———, *The British Army in Ulster*, Vol. 4, Belfast, Northern Ireland: Century Books, 1981.

BBC Monitoring Middle East (Al-Jazeera), "Hamas Says Developing New Weapons to Attack Israel," January 26, 2003.

BBC Monitoring Middle East (Al-Jazeera), "Islamic Jihad Official Interviewed on Strategy of Attacks," July 8, 2004.

BBC News, "Country Profile: Israel and Palestinian Territories," undated. As of December 27, 2006:
http://news.bbc.co.uk/2/hi/middle_east/country_profiles/803257.stm

Beitler, Ruth, *The Path to Mass Rebellion: An Analysis of Two Intifadas*, Lanham, Md.: Lexington Books, 2004.

Bell, Lt. Chuck, JOC Melinda Larson, and JO2 Stephen Haynes, "Philippine and US Forces Train to Fight Terrorism," *Asia-Pacific Defense Forum*, Winter 2004–2005, p. 36.

Bell, J. Bowyer, *On Revolt: Strategies of National Liberation*, Cambridge, Mass.: Harvard University Press, 1976.

———, *IRA: Tactics and Targets*, Swords, Ireland: Poolbeg, 1993.

———, *The Secret Army: The IRA*, Dublin: Poolbeg, 1998.

Ben-David, Alon, "IDF Wary After Qassam 2 Strike," *Jane's Defence Weekly*, September 10, 2003.

————, "Country Briefing—Israel—Double Jeopardy," *Jane's Defence Weekly*, November 17, 2004a.

————, "Gaza: The Ghost of Lebanon," *Jane's Defence Weekly*, May 26, 2004b.

————, "Israel Reactivates Mossad," *Jane's Intelligence Review*, January 2004c.

————, "Israel's Low-Intensity Conflict Doctrine—Inner Conflict," *Jane's Defence Weekly*, September 1, 2004d.

————, "IDF Adapts Doctrine and Structure in Response to Shifting Regional Priorities," *International Defence Review*, March 1, 2005.

Ben-David, Ami, "The Temple Mount as a Fortified Zone," *Ma'ariv*, June 19, 2005, Foreign Broadcast Information Service translation.

Ben-Yisra'el, Major General Yitzhaq, "Impact of Israel's Policy of Targeted Killing on Drop in Palestinian Terror," paper presented to the Tel Aviv University Workshop on Terror and Homeland Security, Tel Aviv, Israel, February 2005, Foreign Broadcast Information Service translation from Hebrew.

Beyler, Clara, "Female Suicide Bombers: Messengers of Death," February 12, 2003. As of July 1, 2005:
http://www.ict.org.il/articles/articledet.cfm?articleid=470

Blanche, Ed, "Palestinian Rocket Finally Hits Large Israeli City—But Only Just," *Jane's Missiles and Rockets*, October 1, 2003.

————, "Stymied Intifada Bombers Turn to Politics," *Jane's Islamic Affairs Analyst*, August 4, 2004.

"Blasts Hit Airport, Hotel and Carrefour," *The Nation*, April 4, 2005, pp. 1–2.

Bonnet, Guillaume, "Israelis Have 400 Eyes in Jerusalem's Old City," *Agence France Presse*, April 3, 2001.

Canadian Security Intelligence Service, "Irish Nationalist Terrorism Outside Ireland: Out-of-Theatre Operations 1972–1993," 1994. As of January 20, 2005:
http://www.csis-scrs.gc.ca/en/publications/commentary/com40.asp

Catignani, Sergio, "The Security Imperative in Counterterror Operations: The Israeli Fight Against Suicidal Terror," *Terrorism and Political Violence*, Vol. 17, No. 1/2, 2005, pp. 245–264.

Chalk, Peter, and Bruce Hoffman, unpublished research on suicide terrorism, 2005.

Chalk, Peter, and William Rosenau, *Confronting "the Enemy Within": Security Intelligence, the Police, and Counterterrorism in Four Democracies*, Santa Monica, Calif.: RAND Corporation, MG-100-RC, 2004. As of December 29, 2006:
http://www.rand.org/pubs/monographs/MG100/

Chow, James S., James Chiesa, Paul Dreyer, Mel Eisman, Theodore W. Karasik, Joel Kvitky, Sherrill Lingel, David Ochmanek, and Chad Shirley, *Protecting Commercial Aviation Against the Shoulder-Fired Missile Threat*, Santa Monica, Calif.: RAND Corporation, OP-106-RC, 2005. As of August 25, 2005: http://www.rand.org/pubs/occasional_papers/OP106/

Coaffee, Jon, "Rings of Steel, Rings of Concrete and Rings of Confidence: Designing Out Terrorism in Central London Pre and Post September 11th," *International Journal of Urban and Regional Research*, Vol. 28, No. 1, 2004, pp. 201–211.

Cobban, Helena, *The Palestinian Liberation Organisation: People, Power, and Politics*, Cambridge, UK, and New York: Cambridge University Press, 1984.

Collins, Eamon, and Mick McGovern, *Killing Rage*, London: Granta Books, 1998.

Commission on Physical Sciences, Mathematics, and Applications, *Containing the Threat from Illegal Bombings: An Integrated National Strategy for Marking, Tagging, Rendering Inert, and Licensing Explosives and Their Precursors*, Washington, D.C.: National Academies, 1988.

Coogan, Tim Pat, *The IRA: A History*, Niwot, Colo.: Roberts Rinehart, 1993.

Copans, Laurie, "Israel to Install Biometric ID System at Gaza Checkpoint," Associated Press, December 2, 2003.

"Country on Security Alert," *The Nation*, April 5, 2005, pp. 1–4.

Cragin, Kim, "Hizballah," in Brian A. Jackson, John C. Baker, Peter Chalk, Kim Cragin, John V. Parachini, and Horacio R. Trujillo, *Aptitude for Destruction*, Vol. 2, *Case Studies of Organizational Learning in Five Terrorist Groups*, Santa Monica, Calif.: RAND Corporation, MG-332-NIJ, 2005, pp. 37–56. As of December 27, 2006: http://www.rand.org/pubs/monographs/MG332/index.html

Cragin, R. Kim, Peter Chalk, Sara A. Daly, and Brian A. Jackson, *Sharing the Dragon's Teeth: Terrorist Groups and the Exchange of New Technologies*, Santa Monica, Calif.: RAND Corporation, forthcoming.

Cragin, Kim, and Sara A. Daly, *The Dynamic Terrorist Threat: An Assessment of Group Motivations and Capabilities in a Changing World*, Santa Monica, Calif.: RAND Corporation, MR-1782-AF, 2004. As of December 29, 2006: http://www.rand.org/pubs/monograph_reports/MR1782/

Davis, Anthony, "Tracking Tigers in Phuket," *Asiaweek*, June 16, 2000, p. 28.

———, "Tamil Tigers Seek to Rebuild Naval Force," *Jane's Intelligence Review*, March 1, 2005a.

———, "Thai Militants Adopt New Bombing Tactics," *Jane's Intelligence Review*, Vol. 17, No. 5, May 1, 2005b, pp. 26–31.

de Silva, K., *Sri Lanka: Ethnic Conflict, Management and Resolution*, Kandy, Sri Lanka: International Centre for Ethnic Studies, 1996, p. 4.

Dewar, Lt-Col Michael, *The British Army in Northern Ireland*, London: Arms and Armour Press, 1985.

Dillon, Martin, *The Dirty War*, London: Arrow Books, 1990.

Dixit, J., "Indian Involvement in Sri Lanka and the Indo-Sri Lankan Agreement of 1987: A Retrospective Evaluation," in Kumar Rupesinghe, ed., *Negotiating Peace in Sri Lanka: Efforts, Failures and Lessons*, London: International Alert, 1998.

Dolnik, Adam, and Anjali Bhattacharjee, "Hamas: Suicide Bombings, Rockets, or WMD?" *Terrorism and Political Violence*, Vol. 14, No. 3, 2002, pp. 109–128.

Drake, C. J. M., "The Provisional IRA: A Case Study," *Terrorism and Political Violence*, Vol. 3, No. 2, 1991, pp. 43–60.

Dudkevitch, Margot, "Terrorist Describes Sending Bombers to Their Targets," *Jerusalem Post*, June 12, 2002, p. 3.

———, "Terrorists Step up Arms Smuggling, 'Post' Learns," *Jerusalem Post*, March 31, 2005.

Elon, Emuna, "Beyond the Illusion of the Fence," *Jerusalem Report*, August 26, 2002.

Enders, Walter, and Todd Sandler, "What Do We Know About the Substitution Effect in Transnational Terrorism?" in Andrew Silke, ed., *Research on Terrorism: Trends, Achievements and Failures*, London: Frank Cass, 2004, pp. 119–137.

Eshel, David, "Israel Refines Its Pre-Emptive Approach to Counterterrorism," *Jane's Intelligence Review*, September 2002a.

———, "Israel Hones Intelligence Operations to Counter Intifada," *Jane's Intelligence Review*, October 2002b.

Esposito, Michele, "The al-Aqsa Intifada: Military Operations, Suicide Attacks, Assassinations, and Losses in the First Four Years," *Journal of Palestinian Studies*, Vol. 34, No. 2, 2005, pp. 85–122.

The Europa Yearbook: A World Survey, London: Europa Publications, 1998.

"Expert Warns of Breast Type Suicide Kits," *ColomboPage News Desk*, January 10, 2005.

"Female Bomber a Hamas First," *Toronto Star*, January 15, 2004, p. A10.

Fighel, Yoni, *Hamas' Next Strategic Weapon in the West Bank*, Herzlia, Israel: Institute for Counter-Terrorism, 2005a.

———, "The Qassam Rockets: Hamas' Next Strategic Weapon in the West Bank," July 13, 2005b. As of December 29, 2006: http://www.ict.org.il/apage/5287.php

Fishman, Alex, "The Revenge of the D-4," *Yedi'ot Aharonot*, May 21, 2004a, Foreign Broadcast Information Service translated text.

———, "The Partition Plan," *Yedi'ot Aharonot*, December 3, 2004b, Foreign Broadcast Information Service translation.

———, "An End to the Era of Bulldozers," *Yedi'ot Aharonot*, December 17, 2004c, Foreign Broadcast Information Service translation.

———, "Putting Out the Fire, Gaining Time," *Yedi'ot Aharonot*, January 21, 2005a, Foreign Broadcast Information Service translation.

———, "The Intifadah Will Resume Next Fall," *Yedi'ot Aharonot*, April 1, 2005b, Foreign Broadcast Information Service translation.

———, "An Iron Fist at the End of Summer," *Yedi'ot Aharonot*, May 6, 2005c, Foreign Broadcast Information Service translation.

"Five Days in an IRA Training Camp," *Iris*, No. 7, November 1983, pp. 39–45.

Foulger, Brian, and Peter J. Hubbard, "A Review of Techniques Examined by U.K. Authorities to Prevent or Inhibit the Illegal Use of Fertiliser in Terrorist Devices," International Explosives Symposium, Fairfax, Virginia, April 1996.

Frisch, Felix. "IDF: There Is a Solution to the Problem of the Tunnels," *Tel Aviv Ma'ariv*, May 23, 2005, Foreign Broadcast Information Service translation.

Geraghty, Tony, *The Irish War: The Hidden Conflict Between the IRA and British Intelligence*, Baltimore, Md.: Johns Hopkins University Press, 2000.

Gilmour, Raymond, *Dead Ground: Infiltrating the IRA*, London: Little, Brown, 1998.

Globalsecurity.org, "Jemaah Islamiya," undated. As of October 11, 2005: http://www.globalsecurity.org/military/world/para/ji.htm

Greenlees, Don, "Still a Force to Be Feared," *Far Eastern Economic Review*, January 22, 2004, pp. 14–17.

Grinberg, Hanan, "IDF Building New Gaza Barrier," *Tel Aviv Ynetnews*, April 14, 2005, Foreign Broadcast Information Service transcription.

Gunaratna, Rohan, *International and Regional Security Implications of the Sri Lankan Tamil Insurgency*, Colombo, Sri Lanka: Alumni Association of the Bandaranaike Centre for International Studies, 1997.

———, *Sri Lanka's Ethnic Crisis and National Security*, Colombo, Sri Lanka: South Asian Network on Conflict Research, 1998.

———, "Suicide Terrorism in Sri Lanka and India: LTTE Suicide Capability, Likely Trends and Response," unpublished paper, 2000.

————, *Inside Al Qaeda: Global Network of Terror*, New York: Columbia University Press, 2002.

————, "Understanding Al Qaeda and Its Network in Southeast Asia," in Kumar Ramakrishna and See Seng Tan, eds., *After Bali: The Threat of Terrorism in Southeast Asia*, Singapore: Institute of Defense and Strategic Studies, Nanyang Technological University, 2004, pp. 117–132.

"HAMAS Rep in Lebanon on Killing HAMAS Member in Syria, Resistance, Other Issues," *Al-Mustaqbal* (online edition), October 1, 2004, Foreign Broadcast Information Service translation.

Hamill, Desmond, *Pig in the Middle: The Army in Northern Ireland, 1969–1984*, London: Methuen, 1985.

Harel, Amos, "IDF Plans to Build Trench along Philadelphi Road," *Haaretz*, April 28, 2004a.

————, "Gaza Terror Groups Thought to Have Anti-Aircraft Missiles," *Haaretz*, October 25, 2004b.

————, "5 Soldiers Killed in Rafah Tunnel Attack," *Haaretz*, December 13, 2004c.

————, " Security Forces Fear Bombers May Learn from Car Thieves How to Get Around the Separation Fence," *Haaretz*, January 17, 2005a.

————, "IDF: Defense of Towns Along W. Bank Border Not Our First Priority," *Haaretz*, June 1, 2005b.

Harnden, Toby, *Bandit Country: The IRA and South Armagh*, London: Coronet Books, LIR, 2000.

Hausman, Tamar, "Police Surveillance Cameras to Keep Peace in Jerusalem's Old City," *Jerusalem Post*, January 10, 2000.

Hewett, Jennifer, "There Are Many, Many More Victims," *Sydney Morning Herald*, October 15, 2002, p. A1.

Holland, Jack, and Susan Phoenix, *Phoenix: Policing the Shadows*, London: Hodder and Stoughton, 1996.

Horgan, John, and Max Taylor, "The Provisional Irish Republican Army: Command and Functional Structure," *Terrorism and Political Violence*, Vol. 9, No. 3, 1997, pp. 1–32.

————, "Playing the 'Green Card'—Financing the Provisional IRA: Part 1," *Terrorism and Political Violence*, Vol. 11, No. 2, Summer 1999, pp. 1–38.

————, "Playing the 'Green Card'—Financing the Provisional IRA: Part 2," *Terrorism and Political Violence*, Vol. 15, No. 2, Summer 2003, pp. 1–60.

Harub, Khalid, *Hamas: Political Thought and Practice*, Washington, D.C.: Institute for Palestine Studies, 2000.

————, "Hamas After Shaykh Yasin and Rantisi," *Journal of Palestinian Studies*, Vol. 34, No. 4, 2004, pp. 21–38.

IDF—*see* Israel Defense Forces.

Independent Monitoring Commission, Northern Ireland Office, *First Report of the Independent Monitoring Commission: Presented to the Government of the United Kingdom and the Government of Ireland Under Articles 4 and 7 of the International Agreement Establishing the Independent Monitoring Commission*, London: Government of the United Kingdom, Her Majesty's Stationery Office, 2004.

International Crisis Group, *Indonesia Backgrounder: How the Jemaah Islamiyah Terrorist Network Operates*, Asia Report No. 43, December 11, 2002.

————, *Jemaah Islamiyah in South East Asia: Damaged but Still Dangerous*, Asia Report No. 63, August 26, 2003.

————, *Southern Philippines Backgrounder: Terrorism and the Peace Process*, Asia Report No. 80, July 13, 2004.

International Institute for Strategic Studies, "Sri Lanka's Peace Process in Jeopardy," *Strategic Comments*, Vol. 10, No. 3, April 2004.

International Policy Institute for Counter Terrorism, "Hamas: Islamic Resistance Movement," undated. As of July 1, 2005:
http://www.ict.org.il/organizations/orgdet.cfm?orgid=13

"IRA 'Has Destroyed All Its Arms,'" BBC, September 26, 2005. As of November 17, 2005:
http://news.bbc.co.uk/1/hi/northern_ireland/4283444.stm

"IRA Interview: Iris Talks to a Member of the IRA's General Headquarters Staff," *Iris*, Vol. 1, No. 1, April 1981, pp. 42–48.

"ISA Arrests Head of Gaza Strip Hezbollah Cell," Israel Ministry of Foreign Affairs, March 10, 2004. As of December 28, 2006:
http://www.mfa.gov.il/MFA/Terrorism-+Obstacle+to+Peace/Terrorism+and+Islamic+Fundamentalism-/ISA+arrests+Gaza+Hezbollah+cell+10-Mar-2004.htm

Israel Defense Forces, "Details of Security Forces and Israeli Civilians Killed Since September 2000," undated. As of June 27, 2005:
http://www1.idf.il/SIP_STORAGE/DOVER/files/4/33524.doc

————, "The End of Manual Security Checks," September 10, 2004, Foreign Broadcast Information Service translation.

Israel Ministry of Defense, "Israel's Security Fence: Route," February 20, 2005. As of November 22, 2005:
http://www.seamzone.mod.gov.il/Pages/ENG/route.htm

Jackson, Brian A., "Technology Acquisition by Terrorist Groups: Threat Assessment Informed by Lessons from Private Sector Technology Adoption," *Studies in Conflict and Terrorism*, Vol. 24, 2001, pp. 183–213.

———, "The Provisional Irish Republican Army," in Brian A. Jackson, John C. Baker, Peter Chalk, Kim Cragin, John V. Parachini, and Horacio R. Trujillo, *Aptitude for Destruction*, Vol. 2: *Case Studies of Learning in Five Terrorist Organizations*, Santa Monica, Calif.: RAND Corporation, MG-332-NIJ, 2005, pp. 93–140. As of December 28, 2006:
http://www.rand.org/pubs/monographs/MG332/

Jackson, Brian A., John C. Baker, Peter Chalk, Kim Cragin, John V. Parachini, and Horacio R. Trujillo, *Aptitude for Destruction*, Vol. 1: *Organizational Learning in Terrorist Groups and Its Implications for Combating Terrorism*, Santa Monica, Calif.: RAND Corporation, MG-331-NIJ, 2005a. As of December 28, 2006:
http://www.rand.org/pubs/monographs/MG331/

———, *Aptitude for Destruction*, Vol. 2: *Case Studies of Learning in Five Terrorist Organizations*, Santa Monica, Calif.: RAND Corporation, MG-332-NIJ, 2005b. As of December 28, 2006:
http://www.rand.org/pubs/monographs/MG332/

Jane's Foreign Report, "An Eavesdropper's Paradise," March 7, 2002.

Jane's Intelligence Digest, "Hamas' New Terror Tactics?" May 9, 2003.

Jane's Sentinel Security Assessment—Eastern Mediterranean, March 1, 2005.

Jane's World Insurgency and Terrorism, "Provisional Irish Republican Army (PIRA)," 2004.

Jayasinghe, Amal, "Tiger Bombers Primed for a Repeat," *The Australian*, February 8, 1996.

Jenkins, Brian Michael, and Larry N. Gersten, *Protecting Public Surface Transportation Against Terrorism and Serious Crime: Continuing Research on Best Security Practices*, San Jose, Calif.: Mineta Transportation Institute, 2001.

"JI Militants Said Plotting 'Major' Attacks on US, British Embassies," *Philippine Star*, August 13, 2005.

Jones, Sidney, "Jakarta and Jihad: Indonesia Faces More Terror," *International Herald Tribune*, August 29, 2003. As of December 28, 2006:
http://www.iht.com/articles/2003/08/29/edjones_ed3_.php

———, "The Fifth Column: What Indonesia Must Explain," *Far Eastern Economic Review*, September 23, 2004.

Joshi, Manoj, "On the Razor's Edge: The Liberation Tigers of Tamil Eelam," *Studies in Conflict and Terrorism*, Vol. 19, 1996, pp. 19–42.

Karmon, Ely, "Hamas' Terrorism Strategy: Operational Limitations and Political Constraints," *Middle East Review of International Affairs*, March, 2000, pp. 66–79.

Katz, Samuel M., *The Hunt for the Engineer: How Israeli Agents Tracked the Hamas Master Bomber*, New York: Fromm International, 1999.

"Lanka Suspects Submarine in Thailand to be LTTE's," *Associated Press*, July 16, 2000.

Lefkovits, Etgar, "Police Issue Guide to Identifying Bombers," *Jerusalem Post*, August 29, 2002, p. 3.

Lobe, Jim, "Israeli Trench Is Pretext for Uprooting Arabs, HRW Says," *Inter-Press Service*, January 24, 2005.

"LTTE Suicide Kit Assembly Plant in Dehiwala Raided," *The Island* (Sri Lanka), March 14, 2001.

Macintyre, Donald, "Israeli Destruction of Palestinian Homes Violates International Law," *The Independent* (London), October 19, 2004.

MacStiofáin, Seán, *Memoirs of a Revolutionary*, Edinburgh: R&R Clark, 1975.

Marine Corps Intelligence Activity, "Urban Warfare Study: City Case Studies Compilation," 1999. As of March 22, 2005:
http://www.smallwarsjournal.com/documents/urbancasestudies.pdf

Marques, Maj. Patrick D., *Guerilla Warfare Tactics in Urban Environments*, Masters, Fort Leavenworth, Kans.: U.S. Army Command and General Staff College, 2003.

McBeth, John, "Bombs, the Army and Suharto," *Far Eastern Economic Review*, February 1, 2001.

McCorkell, W. J., and R. M. E. Griffin, "An Overview of the Scientific Examinations Performed After an Explosion on the Shankill Road," *Science and Justice*, Vol. 38, 1998, pp. 75–79.

McGartland, Martin, *Fifty Dead Men Walking*, London: Blake Publishing, 1997.

McGreal, Chris, "Informer in Pay of Israel Unbowed by Brother's Bloody Fate," *The Guardian*, August 4, 2004. As of November 22, 2005:
http://www.guardian.co.uk/israel/Story/0,2763,1275360,00.html

Meijer, Roel, *Inventory of the Collection of al-Qiyada al-Wataniyya al-Muwahhida li-l-Intifada (Unified National Command of the Intifada) 1987–1990*, Amsterdam: International Institute of Social History, 1998. As of December 29, 2006:
http://www.iisg.nl/publications/intifada.pdf

Middle East Newsline, "Israeli Security Fence Reported to Lack UAV, Ground Sensor Capabilities," April 20, 2004, Foreign Broadcast Information Service translation.

Ministry of Finance and Planning, *Statistical Pocketbook of the Democratic Socialist Republic of Sri Lanka—1998*, Colombo, Sri Lanka: Department of Census and Statistics, 1998.

Mishal, Shaul, and Avraham Sela, *The Palestinian Hamas: Vision, Violence, and Coexistence*, New York: Columbia University Press, 2000.

Mitchell, Richard P., *The Society of the Muslim Brothers*, London: Oxford University Press, 1969.

Moghadam, Assaf, "Palestinian Suicide Terrorism in the Second Intifada: Motivations and Organizational Aspects," *Studies in Conflict and Terrorism*, Vol. 26, No. 2, 2003, pp. 65–92.

Moore, Molly. "Refuge Is Prison for Hunted Palestinian," *The Washington Post*, August 23, 2004, p. 1.

Morgenstern, Joseph, "Biometrics Help Secure Israel's Borders," *Globes* (online edition), October 9, 2003.

Morris Tribunal, *Report of the Tribunal of Inquiry: Set up Pursuant to the Tribunal of Inquiry (Evidence) Acts 1921–2002 into Certain Gardaí in the Donegal Division*, Dublin: Government of Ireland, 2004.

Murphy, Jr., John F., "The IRA and the FARC in Colombia," *International Journal of Intelligence and Counterintelligence*, Vol. 18, 2005, pp. 76–88.

Nakashima, Ellen, "Militant Convicted in Bali Bombings," *The Washington Post*, August 8, 2003, p. A13.

Nakashima, Ellen, and Alan Sipress, "Tips, Traced Call Led to Capture of Al Qaeda Suspect: Hambali Tracked Through 4 Countries," *The Washington Post*, August 16, 2003, p. A14.

Nassar, Jamal R., and Roger Heacock, eds., *Intifada: Palestine at the Crossroads*, New York: Praeger, 1990.

National Commission on Terrorist Attacks upon the United States, *The 9/11 Commission Report: Final Report of the National Commission on Terrorist Attacks upon the United States*, New York: W. W. Norton, 2004.

National Memorial Institute for the Prevention of Terrorism, *MIPT Terrorism Knowledge Base*, undated database. As of December 28, 2006:
http://www.tkb.org

Netline Communications Technologies Ltd., undated homepage. As of November 21, 2005:
http://www.netline.co.il/Netline/

O'Ballance, Edgar, *Terror in Ireland: The Heritage of Hate*, Novato, Calif.: Presido Press, 1981.

O'Callaghan, Sean, *The Informer*, London: Corgi Books, 1999.

O'Sullivan, Arieh, "Mofaz Calls Completed First Stage of Security Fence 'Impenetrable,'" *Jerusalem Post*, August 20, 2003, Foreign Broadcast Information Service translation.

O'Sullivan, Arieh, and Khaled Abu Toameh, "IDF Relying on Use of Killer Drones in Gaza," *Jerusalem Post*, October 26, 2004.

Palestinian Academic Society for the Study of International Affairs (PASSIA), *The Phenomenon of Collaborators in Palestine*, Jerusalem: PASSIA, 2001.

Palestinian Ministry of Foreign Affairs, "Statistics," undated Web page. As of January 19, 2006:
http://www.mofa.gov.ps/Statistics/index.asp

PASSIA—*see* Palestinian Academic Society for the Study of International Affairs.

Patrick, Derrick, *Fetch Felix: The Fight Against the Ulster Bombers 1976–1977*, London: Hamish Hamilton, 1981.

PBS Frontline, "Transcript: Battle for the Holy Land," April 4, 2002. As of November 21, 2005:
http://www.pbs.org/wgbh/pages/frontline/shows/holy/etc/script.html

"Peace Process Bogged Down in More Questions," *Sunday Times* (Sri Lanka), May 2, 2004.

"Peace Talks: LTTE Not Likely to Respond Soon," *Sunday Times* (Sri Lanka), April 25, 2004.

Quinn, C. C., "The Role of the Forensic Scientist at Pseudo-Military Incidents," *Science and Justice*, Vol. 38, 1998, pp. 85–92.

Rabasa, Angel M., *Political Islam in Southeast Asia: Moderates, Radicals and Terrorists*, London: International Institute for Strategic Studies, Adelphi Paper No. 358, May 2003.

Ramakrishna, Kumar, "US Strategy in Southeast Asia: Counter-Terrorist or Counter-Terrorism?" in Kumar Ramakrishna and See Seng Tan, eds., *After Bali: The Threat of Terrorism in Southeast Asia*, Singapore: Institute of Defense and Strategic Studies, Nanyang Technological University, 2004, pp. 305–340.

Reeves, Phil, "Palestinian Disguised as Soldier Guns Down Three," *The Independent* (London), October 5, 2001.

Ressa, Maria A., *Seeds of Terror: An Eyewitness Account of Al-Qaeda's Newest Center of Operations in Southeast Asia*, New York: Free Press, 2003.

Richardson, Doug, "IDF Hunts Qassam-II Rocket Workshops," *Jane's Missiles and Rockets*, April 1, 2002.

———, "Qassam Range Now 14 km, Says Hamas," *Jane's Missiles and Rockets*, December 1, 2004.

———, "Qassam Rockets Tested in the West Bank," *Jane's Missiles and Rockets*, June 1, 2005.

Rodan, Steve, "Israel Launches 'Virtual Fence' Project for Border Security," *Jane's Defense Weekly*, April 3, 2002.

Ryder, Chris, *The Ulster Defense Regiment: An Instrument of Peace?* London: Methuen, 1991.

———, *The RUC 1922–1997: A Force Under Fire*, London: Mandarin, Random House UK, 1997.

Said, Edward, *The Palestinian Question*, New York: Vintage Books, 1979.

Schechter, Erik, "Where Have All the Bombers Gone?" *Jerusalem Post*, August 6, 2004.

Sebastian, Leonard C., "The Indonesian Dilemma: How to Participate in the War on Terror Without Becoming a National Security State," in Kumar Ramakrishna and See Seng Tan, eds., *After Bali: The Threat of Terrorism in Southeast Asia*, Singapore: Institute of Defense and Strategic Studies, Nanyang Technological University, 2004, pp. 357–382.

Shatz, Adam, "In Search of Hezbollah," *New York Review of Books*, April 29, 2004.

Sinai, Joshua, "Intifada Drives Both Sides to Radical Arms," *Jane's Intelligence Review*, May 1, 2001.

Singapore Ministry of Home Affairs, "The Jemaah Islamiyah Arrests and the Threat of Terrorism," Singapore: Ministry of Home Affairs, January 7, 2003.

Singh, Daljit, "ASEAN Counter-Terror Strategies and Cooperation," in Kumar Ramakrishna and See Seng Tan, eds., *After Bali: The Threat of Terrorism in Southeast Asia*, Singapore: Institute of Defense and Strategic Studies, Nanyang Technological University, 2004, pp. 201–220.

Sipress, Alan, "An Indonesian's Prison Memoir Takes Holy War into Cyberspace in Sign of New Threat, Militant Offers Tips on Credit Card Fraud," *The Washington Post*, December 14, 2004, p. A19.

Sipress, Alan, and Ellen Nakashima, "Jakarta Blast That Killed 9 Blamed on Muslim Extremists," *The Washington Post*, September 10, 2004, p. A18.

Stevenson, Andrew, and Mark Baker, "Volunteers Dig Deep to Help the Dying and Injured," *Melbourne Age*, October 15, 2002, p. A1.

Stevenson, Jonathan, "25 Years of Bloody Conflict Bring Enviable Quality Care; Top Notch Medical Skills Byproduct of N. Ireland's Violence As Hospitals Perform Near Miraculous Feats with Injuries," *Rocky Mountain News*, September 1, 1994, p. 42A.

Styles, George, *Bombs Have No Pity: My War Against Terrorism*, London: William Luscombe, 1975.

Susser, Leslie, "Smugglers Row," *Jerusalem Report*, May 15, 2005.

Tan, Andrew, "The Indigenous Roots of Conflict in Southeast Asia: The Case of Mindanao," in Kumar Ramakrishna and See Seng Tan, eds., *After Bali: The Threat of Terrorism in Southeast Asia*, Singapore: Institute of Defense and Strategic Studies, Nanyang Technological University, 2004, pp. 97–116.

Taylor, Peter, *Brits: The War Against the IRA*, London: Bloomsbury, 2001.

Thackrah, John, *Terrorism and Political Violence*, London: Routledge, 1987.

Thomas, Raju, "Secessionist Movements in South Asia," *Survival*, Vol. 36, No. 2, 1994, pp. 92–114.

"Tighter Security at Airports," *Bangkok Post*, April 5, 2005, p. 1.

Tilakaratna, Berbard, "The Sri Lanka Government and Peace Efforts up to the Indo-Sri Lanka Accord: Lessons and Experiences," in Kumar Rupesinghe, ed., *Negotiating Peace in Sri Lanka: Efforts, Failures and Lessons*, London: International Alert, 1998.

Tortermvasana, Komsan, "Misuse of Mobile Phones Worries Authorities," *Bangkok Post*, April 5, 2005, p. 1.

Trujillo, Horacio R., "The Radical Environmentalist Movement," in Brian A. Jackson, John C. Baker, Peter Chalk, Kim Cragin, John V. Parachini, and Horacio R. Trujillo, *Aptitude for Destruction*, Vol. 2: *Case Studies of Learning in Five Terrorist Organizations*, Santa Monica, Calif.: RAND Corporation, MG-332-NIJ, 2005, pp. 93–140. As of December 28, 2006:
http://www.rand.org/pubs/monographs/MG332/index.html

Turnbull, Wayne, *A Tangled Web of Southeast Asian Islamic Terrorism: The Jemaah Islamiyah Terrorist Network*, Monterey, Calif.: Monterey Institute of International Studies, July 31, 2003. As of May 23, 2005:
http://www.terrorismcentral.com/Library/terroristgroups/JemaahIslamiyah/JITerror/WestJava.html

"Two Die in Triple Hat Yai Blasts," *Bangkok Post*, April 4, 2005, p. 1.

U.S. National Foreign Assessment Center, and Central Intelligence Agency, *The World Factbook*, Washington, D.C.: Central Intelligence Agency, annually since 1981. As of December 27, 2006:
http://purl.access.gpo.gov/GPO/LPS552

Urban, Mark, *Big Boys' Rules: The Secret Struggle Against the IRA*, London: Faber and Faber, 1992.

van Meter, Karl M., "Terrorists/Liberators: Researching and Dealing with Adversary Social Networks," *Connections*, Vol. 24, No. 3, 2002, pp. 66–68.

Vaughn, Bruce, Emma Chanlett-Avery, Richard Cronin, Mark Manyin, and Larry Niksch, *Terrorism in Southeast Asia*, Ft. Belvoir, Va.: Defense Technical

Information Center, 2005. As of December 29, 2006:
http://handle.dtic.mil/100.2/ADA444939

Wijayanta, Hanibal W. Y., "Fireballs from Soap Bars," *Tempo*, November 17, 2003, p. 16–19.

Yapa, Vijitha, "Tamil Lorry Bomb Rips Apart Central Bank in Colombo," *The Independent* (London), February 1, 1996.

Yusuf, Zulkarnaen, "Explosives, Bombs, Terror," *Tempo*, September 8, 2003, p. 16.

Zahhar, Mahmud, "Hamas: Waiting for Secular Nationalism to Self-Destruct," *Journal of Palestine Studies*, Vol. 24, No. 3, Spring 1995, pp. 81–88.